Early Nutrition and Lifestyle Factors

Asim K. Duttaroy • Sanjay Basak

Early Nutrition and Lifestyle Factors

Effects on First Trimester Placenta

Springer

Asim K. Duttaroy
Department of Nutrition
University of Oslo
Oslo, Norway

Sanjay Basak
National Institute of Nutrition
Indian Council of Medical Research
Hyderabad, India

ISBN 978-3-319-38802-1 ISBN 978-3-319-38804-5 (eBook)
DOI 10.1007/978-3-319-38804-5

Library of Congress Control Number: 2016947468

Printed on acid-free paper

This Springer imprint is published by Springer Nature
The registered company is Springer International Publishing AG Switzerland

Preface

The placenta is literally the "tree of life." The derivation of the word placenta comes from Latin for cake (placenta), from Greek for flat, slab-like (plakóenta/plakoúnta), and from German for mother cake (mutterkuchen), all referring to the round, flat appearance of the human placenta. Structurally, the placenta is a hemochorial villous organ. The placenta is the highly specialized organ of pregnancy that supports the normal growth and development of the fetus. Growth and function of the placenta are precisely regulated and coordinated to ensure that the exchange of nutrients and waste products between the maternal and fetal circulatory systems operates at maximal efficiency. After implantation, trophoblast cells proliferate and differentiate along two pathways described as villous and extravillous trophoblast cells. Non-migratory, villous cytotrophoblast cells fuse to form the multinucleated syncytiotrophoblast, which forms the outer epithelial layer of the chorionic villi. It is at the terminal branches of the chorionic villi that the majority of fetal/maternal exchange occurs. Extravillous trophoblast (EVT) cells migrate into the decidua and remodel uterine arteries. This facilitates blood flow to the placenta via dilated, compliant vessels, unresponsive to maternal vasomotor control. However, all these tasks depend on normal vascular development within the placenta itself. Placentation begins with the implantation of the blastocyst; the outermost cells of the blastocyst give rise to the trophoblast, a specialized epithelium that during implantation invades the decidua and the inner myometrium, developing the placenta. The cell columns are formed by a subpopulation of cytotrophoblast cells called extravillous trophoblast that proliferates, invades the decidua and superficial layer of myometrium, and transforms the spiral arteries (a terminal branch of the uterine arteries that reach the endometrial surface). Complete transformation of spiral arteries is required for a successful pregnancy since the transformed spiral arteries become low resistance vessels allowing a normal blood flow to the feto-placental unit.

The mechanisms underlying extravillous trophoblast proliferation and invasion have not been fully established, but it is known that many molecular pathways are involved: (a) cellular interaction systems, whether cell–cell (cadherins) or cell-extracellular matrix (integrins), (b) proteolysis systems such as urokinase plasminogen activator (uPA)/plasminogen activator inhibitor type-2 (PAI-2) and matrix metalloproteinase type 9 (MMP-9)/tissue inhibitor of metalloproteinase-3 (TIMP-3), and (c) growth factors/vascular growth factors such as insulin growth factor II (IGF-2) and its binding protein-1 (IGFBP-1), vascular endothelial growth factor (VEGF) and its receptors (Flt-1), and transforming growth factor beta (TGFβ) and its receptor (endoglin), among others.

In this book, we have collected data related to placentation processes and cells involved, growth factors responsible for placentation, and dietary and environmental factors that may influence the processes. Ultimately, the book describes the impact of improper placentation process on feto-placental growth and adult disease program. We believe that this book will be helpful to all those who are working in these areas of reproductive research.

Oslo, Norway Asim K. Duttaroy
Hyderabad, India Sanjay Basak

Contents

Chapter 1
Placentation as a Predictor of Feto-Placental Outcome: Effects of Early Nutrition

Contents

1.1 Introduction

The placenta transfers nutrients, gases, and waste products between maternal and fetal circulations. Placental transfer capacity is determined by a wide range of factors such as placental surface area and thickness, the abundance of transporters, the gradient of concentrations between both maternal and fetal compartments, placental metabolism and utero-placental flows, and other environmental stimuli. Substances need to cross several cellular layers, among them the trophoblast, through different transport mechanisms such as simple or facilitated diffusion or active transport. Early placentation, which is very critical for optimum placental capacity, depends on the extensive remodeling of the maternal uterine vasculature producing low-resistance blood vessels that facilitate the exchange of nutrients and wastes between the mother and the fetus (Zhong et al. 2010). Proper placental development is critical for fetal growth and development (Myatt 2002). The shallow invasion of the spiral arterioles and the maternal decidual stroma by the extravillous trophoblasts (EVTs) results in poor maternal blood flow to the feto-placental unit. Consequently, this reduces oxygen and nutrients' delivery to the fetus. Concomitantly, ischemia–reperfusion injury may follow inappropriate vasoconstriction of untransformed arteries, increasing oxidative stress, syncytiotrophoblast shedding, and maternal systemic vascular inflammation. All these are important features of preeclampsia (PE). In addition, the compromised

A.K. Duttaroy, S. Basak, *Early Nutrition and Lifestyle Factors*,
DOI 10.1007/978-3-319-38804-5_1

placental growth and development may affect fetal growth and development due to the result of insufficient placental transfer of maternal nutrients such as lipids, glucose, amino acids, minerals, and vitamins. There are several other factors that affect the placental transport function such as interrelationships of maternal food intake, availability of nutrients in the maternal circulation, and ability of the placenta to efficiently transport substrates to the fetal circulation.

During the establishment of pregnancy, a human blastocyst implants into the uterine endometrium to facilitate the formation of a functional placenta. Implantation involves the blastocyst adhering to the uterine luminal epithelium before the primitive syncytiotrophoblast and subsequently specialized cells, the extravillous trophoblast (EVT), invade into the maternal uterine walls and decidua in order to engraft and remodel uterine spiral arteries, creating the placental blood supply at the end of the first trimester. The local microenvironment in decidua is thought to play a key role in regulating trophoblast invasion. Cellular and molecular mechanisms involved in human placental blood vessel formation are now better understood (Pavlov et al. 2003; Demir et al. 2004). Early placentation involves vascular modifications by the cooperative effort of maternal natural killer (NK) cells and invasive placental EVTs with the help of several angiogenic growth factors such as vascular endothelial growth factor (VEGF), angiopoietin-like 4 (ANGPTL4), platelet-derived growth factor (PDGF), fibroblast growth factors (FGFs), placental growth factor (PlGF), and matrix metalloproteinases (MMPs). Vessel formation during human placental development occurs by means of both vasculogenesis and angiogenesis (Burton et al. 2009; Demir and Talpur 2010).

Early placentation depends on the invasive properties of EVTs, and this invasion is important to ensure an adequate maternal blood supply to the placenta as inadequate placental vascular development may compromise the feto-placental growth and development (Mainigi et al. 2014). Overall, the evidence suggests that in a healthy pregnancy almost all cell types in the decidua actively promote EVT invasion and, further, reduced EVT invasion towards the end of the first trimester. Therefore, defects in EVT invasion lead to abnormal placentation and thus can affect pregnancy outcomes. This is exemplified by defects in trophoblast invasion and blood flow in pregnancy complications such as PE (Wang et al. 2012; Myatt 2002). In recent years, it has become clear that poor placentation occurs in a broader range of pregnancy complications than initially thought. In addition to the environmental and dietary factors, defective placentation can also be the result of genetic factors and obesity. Moreover, evolutionary pressures derived from the potential conflicts relationship between the fetus and the mother, which include their response to insults, also play a role in determining the phenotypic expression of disorders of the placental bed.

1.2 Preeclampsia and Placentation

PE is the most frequently encountered medical complication during pregnancy (Kang et al. 2011). PE is not simply de novo onset of hypertension and proteinuria in the last half of pregnancy, but rather a syndrome involving many organs, of which the clinical spectrum ranges from relatively mild to life-threatening situation. The etiology of PE is unknown. A large body of evidence, supported by preclinical models of PE, indicates that abnormal placentation early in pregnancy is an important initial event in the onset of PE (Torry et al. 1998; Myatt 2002). This abnormal placentation stimulates the production of oxidative stress, antiangiogenic factors, and cytokines, resulting in generalized maternal vascular dysfunction and the clinical manifestations of PE (Taki et al. 2012). In the past several years, activation of the endothelin (ET) system has emerged as an important pathway causing PE (Maynard et al. 2010; Makris et al. 2007). During pregnancy, highly invasive EVTs acquire vascular-like properties to remodel uterine spiral arterioles. This creates low-resistance, large-diameter vessels that promote utero-placental blood supply to sustain fetal growth. It is widely accepted that inadequate trophoblast invasion and impaired uterine spiral artery remodeling is an initiating factor in the development of PE. This is thought to impair utero-placental arterial flow and lead to placental oxidative stress. One of the proposed hypotheses for development of PE is an imbalance in angiogenic factors, specifically high levels of sFlt1 and sEng with low levels of free PlGF and VEGF, resulting in the endothelial dysfunction that manifests as the clinical symptoms of PE. Levels of sFlt1 are elevated in serum of patients with PE at the time of clinical expression of the disease and appear to be elevated 2–5 weeks before the clinical onset. Several investigators have shown that the increase in maternal circulating sFlt1 precedes the onset of clinical disease and correlates with disease severity. Maternal sFlt1 levels are elevated even more in severe PE, early-onset PE, and PE complicated by a small for gestational age (SGA) baby. Studies have also shown that the plasma concentrations of free VEGF and PlGF are lower in women with severe PE compared with gestationally age-matched normotensive controls. Again, as is seen with sFlt1, circulating levels of sEng are elevated at the time of clinical disease and weeks before clinical onset of PE.

Women with risk factors for developing PE appear to have elevated levels of sFlt1. Specifically, women with twin pregnancies have circulating levels of sFlt1 that are twice those of women with singleton pregnancies. Levels of sFlt1 are also higher in women with molar pregnancies. Because the sFlt1 gene is located on chromosome 13 and women carrying fetuses with trisomy 13 are at increased risk for PE, circulating levels of sFlt1 are higher and free PlGF lower in patients with a pregnancy complicated by trisomy. The placenta is the major source of sFlt1 and that the levels of mRNA for sFlt1 are upregulated in placentas obtained from preeclamptic patients. Hypoxia upregulates the expression of sFlt1 in primary first trimester placental trophoblast. In vitro studies have shown that sFlt1 causes vasoconstriction and endothelial dysfunction. Recently, it was shown that the ratio of VEGF and PlGF to sFlt1 is lower in explants derived from preeclamptic patients. Conditioned media from normal villous explants induced endothelial cell migration and in vitro tube formation, which was attenuated by exogenous sFlt1, while removal of sFlt1 from preeclamptic conditioned media restored migration and tube formation. These findings suggest that elevated levels of sFlt may be an important determinant.

The clinical symptoms of PE are hypertension, proteinuria, and peripheral and/or cerebral edema. PE is also frequently associated with prematurity and intrauterine growth restriction (IUGR), related to impaired insulin-like growth factor (IGF-1) signaling. Global and placenta-specific gene "knockout" animal studies showed the relative significance of a large number of genes in placental development (Taki et al. 2012). Genetic studies showed that disruption of several transcription factors resulted in impaired placental angiogenesis although the downstream target genes are incompletely understood. Fra1 plays a crucial role in establishing normal vascularization of the placenta. HIF-1β mediates hypoxia-induced transcription of many angiogenic genes in the placenta, including VEGF. PPARγ belongs to a family of ligand-activated transcription factors of the nuclear hormone receptor superfamily, which mainly regulates the expression of genes involved in lipid and energy metabolism (Tobin et al. 2006; Aljada et al. 2008). It is also highly expressed in the trophoblast cells of the rodent labyrinth and in the cytotrophoblasts and syncytiotrophoblasts in human placentas, which is increased at late gestation. PPARγ is a critical transcription factor that regulates placental vascular development. PPARγ-null mice are embryonic lethal between E9.5 and E11.5 due to defects in placental vascularization, highlighting its role in placental vascular development.

Regulation of placental angiogenesis, therefore, could become a novel and powerful way for ensuring positive outcomes for most pregnancies. Latest data suggest that several dietary and environmental factors may also regulate angiogenesis via modulating the expression of angiogenic factors (Spencer et al. 2009; Basak and Duttaroy 2013b). Most of these angiogenic factors (VEGF, ANGPTL4) are upregulated by docosahexaenoic acid, $22{:}6n-3$ (DHA) which is an important omega-3 fatty acid and is a structural component of the plasma membranes (Johnsen et al. 2011; Basak et al. 2013). DHA and other fatty acids including CLA and eicosanoids regulate angiogenesis in placental first trimester cells (Murota et al. 1991; Massaro et al. 2010; Johnsen et al. 2011; Zhang and Daaka 2011; Basak and Duttaroy 2012, 2013a, b; Basak et al. 2013; Moon et al. 2003). The dietary fatty acids stimulated the expression of not only major angiogenic factors, such as VEGF and ANGPTL4, but also FABP-4 and FABP-3 which are known to directly modulate angiogenesis (Basak et al. 2013). Among all the FABPs, FABP-4 appeared to be the potent regulator of angiogenic process mediated by DHA, leptin, or VEGF (Elmasri et al. 2009).

1.3 Placentation and Intrauterine Growth Restriction Babies

IUGR is defined as a failure of the fetus to attain its predetermined growth potential. IUGR is usually a disorder of placental insufficiency where the placenta is not providing sufficient nutrients to the baby to sustain normal growth. However, IUGR can also be caused by other factors, including such as abnormal chromosomes, abnormalities in a cluster of genes or a single gene, epigenetic gene silencing, substance abuse (including smoking), pregnancy at high altitude, starvation, and anemia. Placental functional assay is a very useful measure to directly diagnose the placental basis of IUGR. Fetal growth depends on the interactions of genetic and epigenetic determinants functioning against an environment of maternal, fetal, and placental influences (Gardosi et al. 1992). IUGR is a failure to achieve the growth potential promised by these factors. IUGR manifests as a variable syndrome of suboptimal growth and body disproportions rather than a well-defined etiologic entity. Causes for IUGR are diverse and include aneuploidies, non-aneuploid syndromes, infections, metabolic factors, and placental disorders. IUGR places the fetus and neonate at risk of death or disability in the perinatal period and predisposes the child to a lifelong increased risk for hypertension, cardiovascular disorders, and renal disease, among others. The magnitude of the risk varies depending on the a priori risk of the population studied and the definition applied for the diagnosis of IUGR. A common definition is an estimated fetal weight less than the 10th percentile for gestational age, although a fetal weight <3rd percentile is a better predictor of perinatal mortality. Suboptimal growth in abdominal circumference, increased fetal head to abdominal circumference ratio, and oligohydramnios, individually or in combination, reflect placental dysfunction and portend fetal compromise. Doppler ultrasound refines the predictive value for poor outcomes at a given growth percentile. Diminished fetal arterial and venous Doppler flows in key vascular beds predict worsening fetal acid–base status, and such findings frequently lead to delivery of a markedly premature baby to avoid in utero demise.

Optimal placental development and the ability of the placenta to compensate for stimulus-induced injury are central in promotion of normal fetal growth. Histopathological studies of the placenta in IUGR indicate that abnormalities of the maternal spiral arterioles, dysregulated villous vasculogenesis, and abundant fibrin deposition are characteristics of the injuries associated with this condition. Oxidative stress and complement activation, key pathways that regulate apoptosis in villous trophoblast including increased p53 activity, altered translation of AKT and mTOR proteins, and the stress response of the endoplasmic reticulum are responsible for placental dysfunction. Moreover, trophoblast dysregulation at a subcellular level and loss of functional mass of villous trophoblast via cell death pathways are the key contributors to the suboptimal placental performance associated with IUGR. A better understanding of placental dysfunction in IUGR will lead to targeted therapeutic options for this important clinical condition. Placental protein synthesis inhibition induced by ER stress was recently identified as a component of the pathophysiology of human IUGR. The majority of conditions affecting fetal growth are placental or fetal in origin. Fetal weight is determined by the genetic growth potential, the health of the fetus, the capacity of the mother to supply adequate quality and quantities of substrates required for growth, and the ability of the placenta to transport these nutritional substrates to the fetus. In the majority of these cases, there is diminished maternal utero-placental blood flow, caused by insufficient or incomplete trophoblastic invasion of the spiral arteries in the placental bed. Optimal placental development and the compensatory responses of the placenta to exogenous insults are critical in normal pregnancy. Histopathological studies of the placenta in IUGR indicate that abnormalities of the maternal spiral arterioles, dysregulated villous vasculogenesis, and abundant fibrin deposition are characteristics in IUGR. Whether these changes derive from hypoxia, ischemia/reperfusion, complement activation, or other source, where trophoblast layer of villi reflects excess epithelial injury and ER stress that yield dysregulation at a subcellular level, are not known. The increasing gradations of placental dysfunction that evolve in a pregnancy, as the diagnosis of IUGR evolves as a consequence, ultimately result in limited nutrient transfer and reduced blood flow to the fetus. Knowing that delivery of a suboptimally grown, often preterm baby, with a guarded short-term and long-term prognosis, is the end result of IUGR babies. In order to avoid the IUGR babies, it is important to identify the mechanisms responsible for this disorder so that targeted therapeutic options can be developed.

1.4 Placental Programming of Adult Health and Disease

The placenta controls fetal development and growth through the supply of nutrients, respiratory gases and waste, and hormone synthesis. Adverse maternal conditions affect placental growth and consequently modify placental structure and function and thus the fetal outcome (Fig. 1.1). Nevertheless, the placenta is capable of adapting to its environment to optimize its functions and promote fetal survival. In the case of placental insufficiency, fetal growth could be altered, leading to a higher risk of developing metabolic syndrome in adulthood. Several studies have established an effect of environmental stimuli on placental structure and function. Placental circulation through uterine and umbilical blood flow also contributes to the success of the pregnancy. Placental phenotypes are associated with adult onset cardiovascular disease and can be a predictor of adult health and diseases. Sudden cardiac death is associated with a thin placenta, and heart failure is associated with a small placenta in short mothers. Impaired trophoblast invasion has been implicated in a continuum of complications of pregnancy such as unexplained miscarriage, preeclampsia and IUGR, placental abruption, preterm labor, and stillbirth. Future studies would be directed to understand the impact of periconceptional nutrition on placental phenotype and their association with cardiovascular disease. Table 1.1 summarizes the impact of suboptimal placentation and its effects on fetal outcome

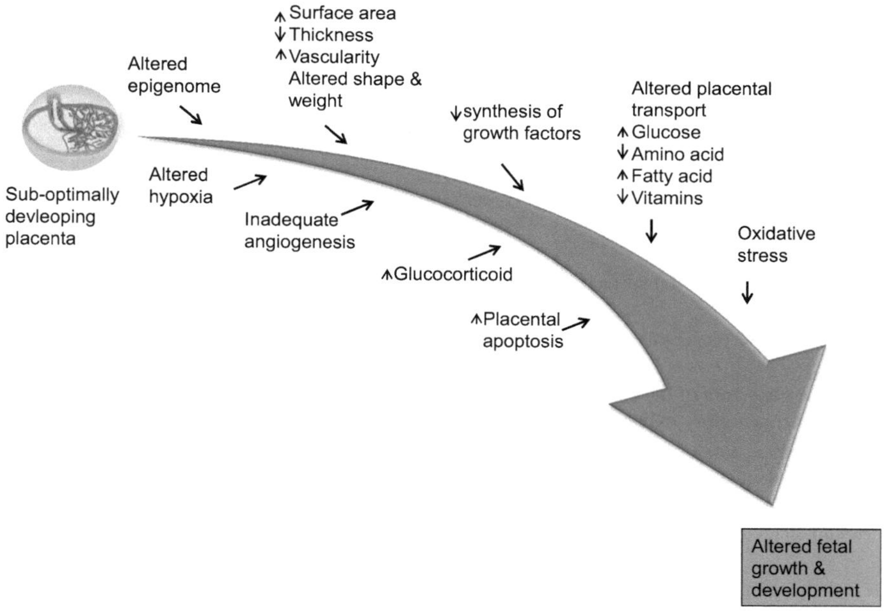

Fig. 1.1 Influence of early placental restriction on fetus growth and development

and risk of adult health and diseases. The long-term effects of the same environmental insult, such as maternal unbalanced nutrition or maternal stress, can have various phenotypic effects on male and female offspring. The placenta has long been considered as a "sexless" organ. Because of its fetal origin, the placenta in fact bears the same genetic information of sex as the fetus, XX or XY, and can also be differentially sensitive to fetal hormones. The majority of studies linking the sexually dimorphic placental response to nutrition and metabolism concern poor maternal nutrition. The Dutch famine at the end of the 2nd World War induced a decrease in placental size and area and a disruption of the fetus-to-placenta ratio index. The effect on the placental area was more severe for boys than girls (Roseboom et al. 2011). Moreover, the health outcomes were also different: in boys, changes in placental shape were associated with the development of the hypertension later in life, which was not the case in girls. Placental formation and function showed important placental defects in mice mutants of DNA methyltransferases. Histone modifications are crucial for establishment of the trophoblast lineage, maintenance of this extraembryonic identity, and placental function (Rugg-Gunn 2012). Altogether, the data presented in this chapter converge on the fact that adaptation in placental phenotype in response to maternal diet and metabolic status (GDM, obesity, undernutrition, etc.) alters fetal nutrient supply, leading to a higher susceptibility to develop metabolic disorders under adverse adult life conditions, such as an obesogenic diet and/or morbidity.

Table 1.1 Placental phenotype as a predictor of health and diseases

Placenta type (size and shape)	Risk predictor of health and disease	References
Low placental weight relation to birth weight	Hypertension in adult	Eriksson et al. (2000)
High placental weight relation to birth weight	Hypertension and type 2 diabetes in adult	Eriksson et al. (2000)
Large placental weight and large placental surface area relation to birth weight	Hypertension in men derived from tall mother with low BMI	Barker et al. (2010)
Breadth of the placental surface	Neonatal body size	Alwasel et al. (2012)
Short length of placental surface	Increase asthma risk	Barker et al. (2013a)
Large surface area of the placenta due to compensatory expansion	Coronary heart disease in men	Eriksson et al. (2011)
Small surface area of the placenta causes inadequate chorionic growth and nutrient transfer	Risk factor for coronary heart disease	Harding (2001)
Round-shaped placental surface	Improved life span	Barker et al. (2011)
Oval-shaped placental surface and reduced breadth of the placenta	– Coronary heart disease in men – Preeclampsia for first pregnancy – Altered placentation	Kajantie et al. (2010), Eriksson et al. (2011)
Increased length of placental surface	Colorectal cancer due to oxidative damage during critical period of organ development	Barker et al. (2013c)
Thinness of the placenta surface	Sudden cardiac death in later life due to impaired development of autonomous nervous system	Barker et al. (2012)
Number of cotyledons in placenta	High blood pressure at age 9 due to altered vascular tone	Barker et al. (2013b)

Sexual dimorphism is another very important aspect of the DOHaD process that is increasingly being taken into account. It is becoming clear that the placenta reacts differently to the same environment depending on the sex of the fetus. The sex specificity of the adult-onset phenotype is, therefore, already partly shaped in utero and the placenta is at stake in this sex-specific feature. Finally, the fetal programming processes and the sex specificities imply important epigenetic mechanisms that are, however, still poorly documented. Effort should be concentrated not only on the observation of sex-specific features but also, and more importantly, on the understanding of the underlying mechanisms. A possible process that may involve the sex-specific epigenome, apposed by epigenetic enzymes encoded by genes on the X-/Y-chromosome, such as Kdm5c/5d, Utx (Kdm6a)/Uty, MeCP2, would be differentially affected by maternal diet and metabolism leading to male and female divergence in programming (please see in Chap. 10). Epigenetic mechanisms, such as DNA methylation, multiple histone modifications, and noncoding RNAs, could contribute in this programming process by governing placental differentiation and function and, therefore, the fetal growth and genotype. Maternal metabolic status can impact epigenetic mark apposition and memory, which could lead to an altered placental gene expression and function with direct consequences for fetal tissue development, and thus potentially have an impact on adult offspring phenotype.

1.5 Summary

Inappropriate placentation may contribute to placental dysfunction and, therefore, this is considered as an important cause of infertility and fetal growth retardation. The increased placental secretion of proinflammatory cytokines and angiogenic regulators is thought to contribute to widespread maternal endothelial dysfunction. The failure of normal placentation due to inappropriate early maternal nutrition generates a series of clinical abnormalities such as PE, IUGR, preterm labor, preterm premature rupture of membranes (PROM), in utero fetal death, and placental abruption as depicted in Fig. 1.2. Emerging evidence also underscores the importance of placental development in long-term health and disease for both the mother and the offspring.

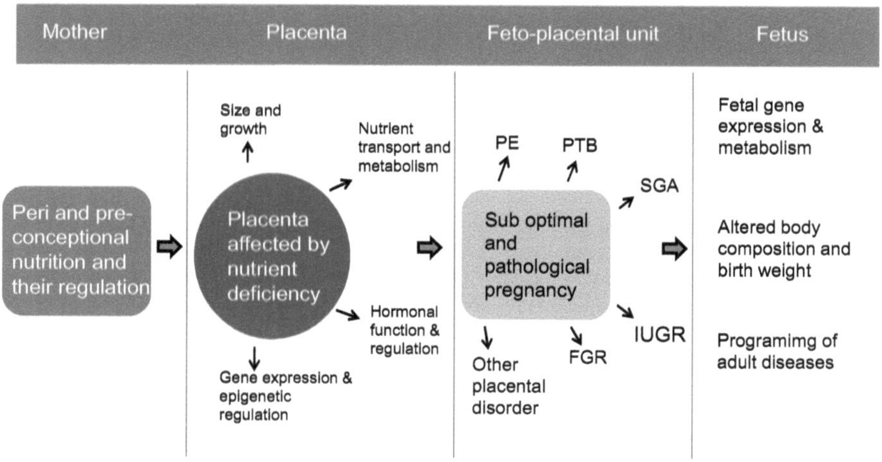

Fig. 1.2 Optimum nutrition influences the development of the placenta affecting early stages of life. *PE* preeclampsia, *PTB* preterm birth, *FGR* fetal growth restriction, *IUGR* intrauterine growth restriction, *SGA* small for gestational age

References

Aljada, A., O'Connor, L., Fu, Y. Y., & Mousa, S. A. (2008). PPAR gamma ligands, rosiglitazone and pioglitazone, inhibit bFGF- and VEGF-mediated angiogenesis. *Angiogenesis, 11*, 361–367.

Alwasel, S. H., Abotalib, Z., Aljarallah, J. S., Osmond, C., Al Omar, S. Y., Harrath, A., Thornburg, K., & Barker, D. J. (2012). The breadth of the placental surface but not the length is associated with body size at birth. *Placenta, 33*, 619–622.

Barker, D. J., Larsen, G., Osmond, C., Thornburg, K. L., Kajantie, E., & Eriksson, J. G. (2012). The placental origins of sudden cardiac death. *International Journal of Epidemiology, 41*, 1394–1399.

Barker, D. J., Osmond, C., Forsen, T. J., Thornburg, K. L., Kajantie, E., & Eriksson, J. G. (2013a). Foetal and childhood growth and asthma in adult life. *Acta Paediatrica, 102*, 732–738.

Barker, D., Osmond, C., Grant, S., Thornburg, K. L., Cooper, C., Ring, S., & Davey-Smith, G. (2013b). Maternal cotyledons at birth predict blood pressure in childhood. *Placenta, 34*, 672–675.

Barker, D. J., Osmond, C., Thornburg, K. L., Kajantie, E., & Eriksson, J. G. (2011). The lifespan of men and the shape of their placental surface at birth. *Placenta, 32*, 783–787.

Barker, D. J., Osmond, C., Thornburg, K. L., Kajantie, E., & Eriksson, J. G. (2013c). The shape of the placental surface at birth and colorectal cancer in later life. *American Journal of Human Biology, 25*, 566–568.

Barker, D. J., Thornburg, K. L., Osmond, C., Kajantie, E., & Eriksson, J. G. (2010). The surface area of the placenta and hypertension in the offspring in later life. *International Journal of Developmental Biology, 54*, 525–530.

Basak, S., Das, M. K., & Duttaroy, A. K. (2013). Fatty acid-induced angiogenesis in first trimester placental trophoblast cells: possible roles of cellular fatty acid-binding proteins. *Life Sciences, 93*, 755–762.

Basak, S., & Duttaroy, A. K. (2012). Leptin induces tube formation in first-trimester extravillous trophoblast cells. *European Journal of Obstetrics, Gynecology, and Reproductive Biology, 164*, 24–29.

Basak, S., & Duttaroy, A. K. (2013a). cis-9, trans-11 conjugated linoleic acid stimulates expression of angiopoietin like-4 in the placental extravillous trophoblast cells. *Biochimica et Biophysica Acta, 1831*, 834–843.

Basak, S., & Duttaroy, A. K. (2013b). Effects of fatty acids on angiogenic activity in the placental extravillious trophoblast cells. *Prostaglandins, Leukotrienes, and Essential Fatty Acids, 88*, 155–162.

Burton, G. J., Charnock-Jones, D. S., & Jauniaux, E. (2009). Regulation of vascular growth and function in the human placenta. *Reproduction, 138*, 895–902.

Demir, R., Kayisli, U. A., Seval, Y., Celik-Ozenci, C., Korgun, E. T., Demir-Weusten, A. Y., & Huppertz, B. (2004). Sequential expression of VEGF and its receptors in human placental villi during very early pregnancy: Differences between placental vasculogenesis and angiogenesis. *Placenta, 25*, 560–572.

Demir, A. S., & Talpur, F. N. (2010). Chemoenzymatic conversion of linoleic acid into conjugated linoleic acid. *Journal of Agricultural and Food Chemistry, 58*, 1646–1652.

Elmasri, H., Karaaslan, C., Teper, Y., Ghelfi, E., Weng, M., Ince, T. A., Kozakewich, H., Bischoff, J., & Cataltepe, S. (2009). Fatty acid binding protein 4 is a target of VEGF and a regulator of cell proliferation in endothelial cells. *The FASEB Journal, 23*, 3865–3873.

Eriksson, J., Forsen, T., Tuomilehto, J., Osmond, C., & Barker, D. (2000). Fetal and childhood growth and hypertension in adult life. *Hypertension, 36*, 790–794.

Eriksson, J. G., Kajantie, E., Thornburg, K. L., Osmond, C., & Barker, D. J. (2011). Mother's body size and placental size predict coronary heart disease in men. *European Heart Journal, 32*, 2297–2303.

Gardosi, J., Chang, A., Kalyan, B., Sahota, D., & Symonds, E. M. (1992). Customised antenatal growth charts. *Lancet, 339*, 283–287.

Harding, J. E. (2001). The nutritional basis of the fetal origins of adult disease. *International Journal of Epidemiology, 30*, 15–23.

Johnsen, G. M., Basak, S., Weedon-Fekjaer, M. S., Staff, A. C., & Duttaroy, A. K. (2011). Docosahexaenoic acid stimulates tube formation in first trimester trophoblast cells, HTR8/SVneo. *Placenta, 32*, 626–632.

Kajantie, E., Thornburg, K. L., Eriksson, J. G., Osmond, C., & Barker, D. J. (2010). In preeclampsia, the placenta grows slowly along its minor axis. *International Journal of Developmental Biology, 54*, 469–473.

Kang, J. H., Song, H., Yoon, J. A., Park, D. Y., Kim, S. H., Lee, K. J., Farina, A., Cho, Y. K., Kim, Y. N., Park, S. W., Kim, G. J., Shim, S. H., & Cha, D. H. (2011). Preeclampsia leads to dysregulation of various signaling pathways in placenta. *Journal of Hypertension, 29*, 928–936.

Mainigi, M. A., Olalere, D., Burd, I., Sapienza, C., Bartolomei, M., & Coutifaris, C. (2014). Peri-implantation hormonal milieu: Elucidating mechanisms of abnormal placentation and fetal growth. *Biology of Reproduction, 90*, 26.

Makris, A., Thornton, C., Thompson, J., Thomson, S., Martin, R., Ogle, R., Waugh, R., Mckenzie, P., Kirwan, P., & Hennessy, A. (2007). Uteroplacental ischemia results in proteinuric hypertension and elevated sFLT-1. *Kidney International, 71*, 977–984.

Massaro, M., Scoditti, E., Carluccio, M. A., Campana, M. C., & DE Caterina, R. (2010). Omega-3 fatty acids, inflammation and angiogenesis: Basic mechanisms behind the cardioprotective effects of fish and fish oils. *Cellular and Molecular Biology, 56*, 59–82.

Maynard, S. E., Moore Simas, T. A., Bur, L., Crawford, S. L., Solitro, M. J., & Meyer, B. A. (2010). Soluble endoglin for the prediction of preeclampsia in a high risk cohort. *Hypertension in Pregnancy, 29*, 330–341.

Moon, E. J., Lee, Y. M., & Kim, K. W. (2003). Anti-angiogenic activity of conjugated linoleic acid on basic fibroblast growth factor-induced angiogenesis. *Oncology Reports, 10*, 617–621.

Murota, S., Kanayasu, T., Nakano-Hayashi, J., & Morita, I. (1991). Involvement of eicosanoids in angiogenesis. *Advances in Prostaglandin, Thromboxane, and Leukotriene Research, 21B*, 623–626.

Myatt, L. (2002). Role of placenta in preeclampsia. *Endocrine, 19*, 103–111.

Pavlov, N., Hatzi, E., Bassaglia, Y., Frendo, J. L., Evain Brion, D., & Badet, J. (2003). Angiogenin distribution in human term placenta, and expression by cultured trophoblastic cells. *Angiogenesis, 6*, 317–330.

Roseboom, T. J., Painter, R. C., de Rooij, S. R., van Abeelen, A. F., Veenendaal, M. V., Osmond, C., & Barker, D. J. (2011). Effects of famine on placental size and efficiency. *Placenta, 32*, 395–399.

Rugg-Gunn, P. J. (2012). Epigenetic features of the mouse trophoblast. *Reproductive Biomedicine Online, 25*, 21–30.

Spencer, L., Mann, C., Metcalfe, M., Webb, M., Pollard, C., Spencer, D., Berry, D., Steward, W., & Dennison, A. (2009). The effect of omega-3 FAs on tumour angiogenesis and their therapeutic potential. *European Journal of Cancer, 45*, 2077–2086.

Taki, A., Abe, M., Komaki, M., Oku, K., Iseki, S., Mizutani, S., & Morita, I. (2012). Expression of angiogenesis-related factors and inflammatory cytokines in placenta and umbilical vessels in pregnancies with preeclampsia and chorioamnionitis/funisitis. *Congenital Anomalies (Kyoto), 52*, 97–103.

Tobin, K. A., Harsem, N. K., Dalen, K. T., Staff, A. C., Nebb, H. I., & Duttaroy, A. K. (2006). Regulation of ADRP expression by long-chain polyunsaturated fatty acids in BeWo cells, a human placental choriocarcinoma cell line. *Journal of Lipid Research, 47*, 815–823.

Torry, D. S., Wang, H. S., Wang, T. H., Caudle, M. R., & Torry, R. J. (1998). Preeclampsia is associated with reduced serum levels of placenta growth factor. *American Journal of Obstetrics and Gynecology, 179*, 1539–1544.

Wang, W., Feng, L., Zhang, H., Hachy, S., Satohisa, S., Laurent, L. C., Parast, M., Zheng, J., & Chen, D. B. (2012). Preeclampsia up-regulates angiogenesis-associated microRNA (i.e., miR-17, -20a, and -20b) that target ephrin-B2 and EPHB4 in human placenta. *Journal of Clinical Endocrinology and Metabolism, 97*, E1051–E1059.

Zhang, Y., & Daaka, Y. (2011). PGE2 promotes angiogenesis through EP4 and PKA Cgamma pathway. *Blood, 118*, 5355–5364.

Zhong, Y., Tuuli, M., & Odibo, A. O. (2010). First-trimester assessment of placenta function and the prediction of preeclampsia and intrauterine growth restriction. *Prenatal Diagnosis, 30*, 293–308.

Chapter 2
Early Placentation Processes

Contents

2.1 Introduction

The process of embryo implantation is a highly sophisticated and synchronized event between activated implanted embryo and a receptive endometrium. Proper development of embryo and implantation of blastocyst largely depends on optimal endometrial receptivity. The process of embryo implantation occurs within restricted period of time known as "window of implantation" which is highly regulated by two prominent hormones such as estrogen (E2) and progesterone (P4). Endometrial receptivity for implantation occur independent of embryo activation. The precise mechanism of each of these hormones on regulation of implantation and endometrium receptivity is not clear. Endometrium is composed of many cell types that include epithelial, endothelial, stromal, and nonresident immune cells. Stromal cells play crucial roles during embryo invasion to the uterine walls.

Because the placenta's primary role is to provide for physiological exchange between the maternal and fetal circulation, the importance of the placental circulation also has long been recognized and is exemplified by the close relationships among fetal weight, placental size, and uterine and umbilical blood flows during normal pregnancies of many mammalian species. The large increase of blood flow

A.K. Duttaroy, S. Basak, *Early Nutrition and Lifestyle Factors*,
DOI 10.1007/978-3-319-38804-5_2

to the uterus during pregnancy, however, results primarily from the formation and growth of the placental vascular bed. Vascular growth, or angiogenesis, is indeed a major component of the increase in placental blood flow throughout gestation. Angiogenesis consists of several basic processes involving biological signals such as hypoxia, ischemia, and/or blood vessel damage which in turn upregulate the expression of pro-angiogenic growth factors that activate their receptors. Consequently, placental angiogenesis is critical to maintain adequate blood flow during gestation, and any alterations in this process can result in an adverse pregnancy. Invasion of extravillous trophoblasts is controlled by many regulatory factors such as angiogenic growth factors, metalloproteinase (MMPs), cytokines, surface integrins, and other Major Histocompatibility Complex (MHC). Also, the differentiation process of trophoblast cells is influenced by various paracrine or autocrine factors. In hemochorial placentation, trophoblast cells of the fetal placenta line the implantation site and invade the maternal decidua, enabling maternal spiral artery remodeling and establishing adequate blood flow between mother and fetus (see Chap. 1). Defects in this process of trophoblast invasion can lead to a number of disorders that have detrimental effects on both the mother and the fetus. Excessive trophoblast invasion may contribute to the pathogenesis of placenta accreta, a severe pregnancy complication in which abnormally deep attachment of the placenta to the myometrium is observed. Shallow invasion results in hypoxic trophoblast tissue, which can cause oxidative stress in the placenta and has been associated with PE and IUGR. Yet, the mechanisms underlying trophoblast invasion are poorly understood. Several genes and transcription factors have been identified as important for regulating this process. Nevertheless, little is known about how the genes are regulated or about the enhancers to which the transcription factors bind to regulate gene expression. Enhancer regions (distal cis-regulatory regions that regulate spatiotemporal gene expression) are crucial to many processes and often contribute to disease when disrupted. Single nucleotide polymorphisms (SNPs) associated with placental disorders often reside in noncoding DNA, indicating the importance of noncoding enhancer elements in these disorders. Next-generation sequencing of mouse and human placentas, as well as computational prediction, is being used to identify putative placenta enhancer elements, but none of these have focused on the process of trophoblast invasion. Identifying genome-wide enhancers that contribute to the process of trophoblast invasion requires assaying placental tissue at an early time point, while invasion is taking place. Most importantly, angiogenic proteins such as VEGF and PlGF are also found to be involved in spiral artery re-modulation. VEGF also inhibits the maturation of antigen-presenting Dendritic Cells (DCs) and, therefore, helps to maintain immature pro-angiogenic DC. These factors are secreted not only by trophoblast cells and villous mesenchymal cells but also by the uterine stromal/glandular myometrial/ endothelial cells and various other immune cells present at maternal–fetal interface. Several factors such as VEGF, ANGPTL4, PDGF, FGF, and PlGF are involved in placental angiogenesis (Aiello and Wong 2000; Folkman and Klagsbrun 1987; Gealekman et al. 2008). Receptors of VEGF, PDGF, and ANGPTLs contain tyrosine kinase activity. Binding of these factors to their specific receptors cause the synthesis of mRNA of these growth factor. In addition, matrix metalloproteinases (MMPs) regulate vascular remodeling and integrity by breaking down the basement membrane. Placental angiogenesis plays a critical role for the establishment of the placental vascularization (Reynolds and Redmer 2001). Angiogenesis is a multistep process that requires the involvement of many biological signals and stimuli regardless of whether it is a physiological or pathological action. Pro-angiogenic factors are activated in response to some physical signals. Blood vessel damage, infarction, and blood flow reduction lead to decreases in oxygen supply. This state is called hypoxia, which is a potent angiogenic trigger that stimulates pro-angiogenic factor activity. Although the underlying mechanisms are not clear, maternal nutrition also likely influences the expression of genes involved in placental development through regulation of various transcription factors. Indeed, several studies demonstrated a role for PPAR in implantation, trophoblast differentiation, and angiogenesis. Alterations in maternal nutrition during pregnancy can affect the expression of PPARs via epigenetic mechanisms or through homocysteine, which is known to compete for PPARs. Growing evidence indicates that suboptimal maternal nutrition can alter placental development. Role of maternal nutrition—particularly micronutrients like folate, vitamin B_{12}, and omega-3 fatty acids—can affect angiogenesis and, thus, placental and fetal growth. Additional animal and human studies need to be undertaken to elucidate the molecular mechanisms through which maternal nutrition regulates PPARs, specifically to determine whether PPARs affect placental angiogenesis directly through angiogenic factors or indirectly by modulating trophoblast differentiation. Regulation of placental angiogenesis, therefore, could become a novel and powerful way for ensuring positive outcomes for most pregnancies. This chapter deals with the factors that regulate placental angiogenesis.

2.2 Angiogenic Factors and Their Regulation

Several pro-angiogenic molecules regulate the synthesis of VEGF and ANGPTL4 in human placental trophoblasts such as tumor necrosis factor alpha (TNF-α), leptin, IL-8, CCN-1, CCN-3, CLA, glucose, DHA, and several other fatty acids. Therefore, these pro-angiogenic molecules will likewise impact angiogenesis in these cells (Ferrara et al. 2003b). Many other molecular pathways are also involved in the early placentation processes such as cellular interaction systems, proteolytic system, insulin growth factor 2 (IGF-2) and IGF-2 binding protein-1, VEGF receptors (Flt-1), transforming growth factor beta (TGFβ), and endoglin (Ferrara et al. 2003a). Neurotrophins such as BDNF and NGF also act as angiogenic factors (Dhobale 2014). VEGF regulates placental angiogenesis via both autocrine and paracrine ways (Alfaidy et al. 2014). VEGF expression increases in the placenta with gestational age. VEGF receptors, VEGFR1 (fms-related tyrosine kinase 1/Flt1), VEGFR2 (kinase insert domain receptor/KDR), and also VEGF co-receptors (neuropilin-1 and -2) are expressed in placental artery endothelial cells. Expression of VEGF, Flt1, and KDR in maternal caruncle and fetal cotyledonary tissues is positively correlated with the placental vascularization and blood flow. This indicates that the VEGF–VEGFR system is critically important in placental angiogenesis (Ferrara et al. 2003b). VEGF regulates all steps involved in the angiogenesis process (Shibuya 2006). VEGF stimulates endothelial expression of proteases such as urokinase-type and tissue-type plasminogen activators and interstitial collagenase that breakdown extracellular matrix and release endothelial cells from anchorage, allowing them to migrate and proliferate (Basak and Duttaroy 2013a). VEGF also recruits pericytes as these cells are required for maturation and stability of the newly formed vessels. Consequently, VEGF has clinical relevance as a diagnostic and/or prognostic marker for several placental development and diseases. VEGF has been proposed as a critical early marker for several placental pathologies including recurrent pregnancy loss, gestational trophoblastic diseases, IUGR, and PE.

In order to transform the spiral arteries, the placental EVTs invade the maternal walls via two different routes: interstitial and endovascular (Lyall et al. 2001). In the interstitial invasion, the EVTs migrate towards the arterial wall replacing the musculoelastic wall. The interstitial trophoblast may also be originated from extravasation of the intraluminal trophoblasts. In the endovascular invasion, the trophoblastic cells infiltrate into the lumens and walls of the arteries replacing the endothelium; the endovascular trophoblast results from intravasation of interstitial trophoblast or by intraluminal migration of trophoblasts (Lyall et al. 2001). During the process of placentation, trophoblast cells attach to the basal membrane surrounding the stroma of these two types of villi (Myatt 2002; Zhong et al. 2010). In the villi, the trophoblast cells fuse to create an external layer called syncytiotrophoblast; at the distal end of the anchoring villi, the EVT breaks the basal membrane and forms cell columns.

2.3 Angiogenic Signaling Pathways

Angiogenic signaling system involves multiple pathways, and these pathways are established in placentation processes. Activation of the MAPK (ERK1/2, JNK1/2, and p38MAPK), PI3K/Akt1, and eNOS/NO pathways is critical for VEGF- and FGF2-stimulated placental angiogenesis (Wang et al. 2009; Hu et al. 2010). Involvement of angiogenic signaling pathways in the placenta is depicted in Fig. 2.1.

2.3.1 The MAPK Pathways

The MAPK pathways are involved in cell proliferation, differentiation, and cell death. Activation of the MAPK pathways is essential for VEGF- and FGF2-

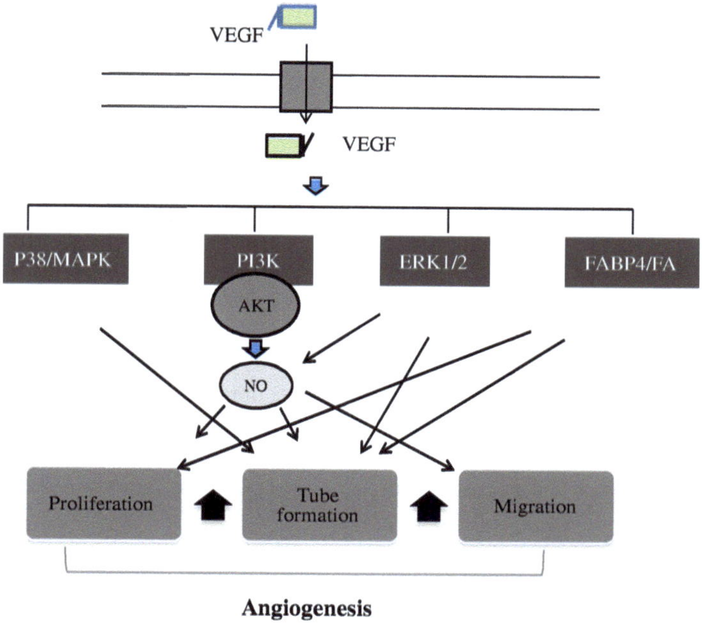

Fig. 2.1 Role of VEGF signaling pathway on angiogenesis of placenta. *FABP4* fatty acid binding protein, *FA* fatty acid

stimulated placental angiogenesis (Zhu et al. 2013; Hu et al. 2010). Several isoforms of MAPK are identified such as the ERKs, p38MAPK, and the JNKs or stress-activated protein kinases. The p38MAPK consists of four members: p38α/MAPK14, p38β/MAPK11, p38γ/MAPK12, and p38δ/MAPK13. Targeted disruption of the p38α gene in mice resulted in embryonic lethality due to severe placental defects (Mudgett et al. 2000) demonstrating the critical roles of P38α in placental development and angiogenesis.

2.3.2 *The PI3K/Akt Pathways*

The family of Akt1 kinases consists of three isoforms: Akt1, Ak2, and AK3 and are encoded by distinct genes. Activation of Akt is initiated by phosphorylation. Activation of Akt1 is involved in VEGF- and FGF2-stimulated eNOS activation. AKt1 is involved in cell proliferation and migration as well as tube formation (Valdes and Corthorn 2011). Akt1-null mouse placenta shows hypotrophy with a decrease in the decidual basalis and glycogen-containing cells in the spongiotrophoblast (Hu et al. 2010). However, Akt2 and Akt3 are not the major players in placental angiogenesis. Inhibition of p38MAPK blocks FGF2-stimulated cell proliferation and migration, but has no effect on VEGF-stimulated cell proliferation and migration (Hu et al. 2010). Inhibition of JNK1/2 significantly decreased placental vascularization and, thus, affects fetal growth and neonatal mortality.

2.3.3 The Nitric Oxide (NO) Pathway

NO is synthesized during the conversion of L-arginine to L-citrulline by NO synthase (NOS). NO is critically involved in VEGF- and FGF2-stimulated angiogenesis (Wang et al. 2009). Inhibition of NOS resulted in preeclampsia-like symptoms and reduced litter size in rats. In eNOS-null pregnant mice, utero-placental remodeling was disturbed resulting in decreased uterine and placental blood flows due to the reduced vascularization in the placenta. eNOS is, therefore, critical for both vasodilation and angiogenesis. The physiological hypoxia also plays a key role in the modulation of the expression of several angiogenic factors including VEGF and ANGPTL4.

2.4 Expression of Angiogenic Genes in Placental Trophoblasts and Their Regulation by Nutrients

The "knockout" animal studies provided informative evidence as to the relative significance of a large number of genes in placental growth and development (Taki et al. 2012). Disruption of several transcription factors results in impaired placental angiogenesis. Fra1 is involved in vascularization of the placenta. HIF-1β regulates hypoxia-induced transcription of many angiogenic genes in human placenta. PPARγ, a regulator of the expression of genes, is involved in lipid and energy metabolism (Tobin et al. 2006; Aljada et al. 2008). PPARγ is a critical transcription factor that regulates placental vascular development. PPARγ is expressed in the trophoblast cells of the rodent labyrinth and in human placental trophoblasts. PPARγ-null mice are embryonic lethal between E9.5 and E11.5 due to defects in placental vascularization, highlighting its role in placental vascular development.

Several dietary and environmental factors may regulate angiogenesis via modulating the expression of angiogenic factors (Spencer et al. 2009; Basak and Duttaroy 2013b). Most of these angiogenic factors are upregulated by docosahexaenoic acid, 22:6$n-3$ (DHA) which is an important omega-3 fatty acid and is a structural component of the plasma membrane (Johnsen et al. 2011; Basak et al. 2013). DHA and other fatty acids including CLA and eicosanoids regulate angiogenesis in placental first trimester cells (Murota et al. 1991; Massaro et al. 2010; Johnsen et al. 2011; Zhang and Daaka 2011; Basak and Duttaroy 2012, 2013a, b; Basak et al. 2013; Moon et al. 2003). The dietary fatty acids increased the expression of not only major angiogenic factors such as VEGF and ANGPTL4

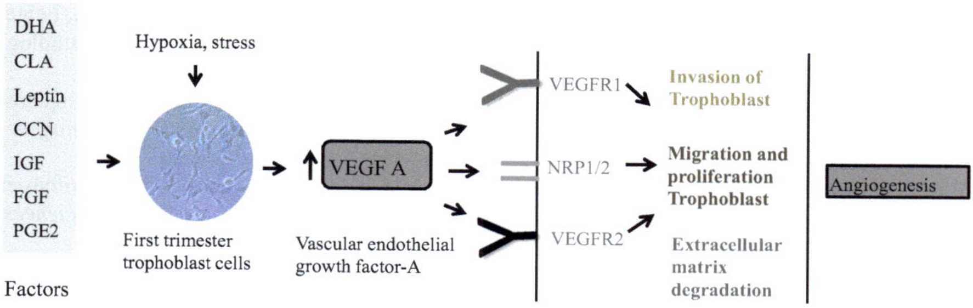

Fig. 2.2 Effects of different factors including dietary factors on VEGF expression and resultant angiogenesis processes in human placenta

but also FABP-4 and FABP-3 which are known to directly modulate angiogenesis (Basak et al. 2013). Among all the FABPs, FABP-4 appeared to be the potent regulator of angiogenic process mediated by DHA, leptin, or VEGF (Elmasri et al. 2009). Figure 2.2 summarizes the effects of dietary and environmental factors controlling angiogenesis via VEGF. Preeclampsia is characterized by an increased placental production of antiangiogenic molecules such as sFlt-1 and sEng which are implicated in the pathogenesis of the disease. Several recent clinical studies suggest that angiogenic biomarkers such as sFlt-1 and PlGF may serve to diagnose and predict preeclampsia and its related complications. In addition, depleting or antagonizing sFlt-1 and sEng in the maternal circulation may prove to be a valuable approach for preeclampsia treatment. Understanding the mechanisms that regulate placental vascular development might allow the development of new preventive strategies for pregnancy complications.

2.5 Summary

Normal placental development is characterized by vasculogenesis and angiogenesis which are tightly regulated by a balance between pro- and antiangiogenic factors. Several angiogenic growth factors and pro-angiogenic factors including certain dietary factors stimulate placental angiogenesis. Abnormalities in this regulation may lead to adverse obstetric outcomes such as preeclampsia and IUGR and also compromise adult health.

References

Aiello, L. P., & Wong, J. S. (2000). Role of vascular endothelial growth factor in diabetic vascular complications. *Kidney International. Supplement, 77*, S113–S119.

Alfaidy, N., Hoffmann, P., Boufettal, H., Samouh, N., Aboussaouira, T., Benharouga, M., Feige, J. J., & Brouillet, S. (2014). The multiple roles of EG-VEGF/PROK1 in normal and pathological placental angiogenesis. *BioMed Research International, 2014*, 451906.

Aljada, A., O'Connor, L., FU, Y. Y., & MOUSA, S. A. (2008). PPAR gamma ligands, rosiglitazone and pioglitazone, inhibit bFGF- and VEGF-mediated angiogenesis. *Angiogenesis, 11*, 361–367.

Basak, S., Das, M. K., & Duttaroy, A. K. (2013). Fatty acid-induced angiogenesis in first trimester placental trophoblast cells: Possible roles of cellular fatty acid-binding proteins. *Life Sciences, 93*, 755–762.

Basak, S., & Duttaroy, A. K. (2012). Leptin induces tube formation in first-trimester extravillous trophoblast cells. *European Journal of Obstetrics, Gynecology, and Reproductive Biology, 164*, 24–29.

Basak, S., & Duttaroy, A. K. (2013a). cis-9, trans-11 conjugated linoleic acid stimulates expression of angiopoietin like-4 in the placental extravillous trophoblast cells. *Biochimica et Biophysica Acta, 1831*, 834–843.

Basak, S., & Duttaroy, A. K. (2013b). Effects of fatty acids on angiogenic activity in the placental extravillious trophoblast cells. *Prostaglandins, Leukotrienes, and Essential Fatty Acids, 88*, 155–162.

Dhobale, M. (2014). Neurotrophins: Role in adverse pregnancy outcome. *International Journal of Developmental Neuroscience, 37C*, 8–14.

Elmasri, H., Karaaslan, C., Teper, Y., Ghelfi, E., Weng, M., Ince, T. A., Kozakewich, H., Bischoff, J., & Cataltepe, S. (2009). Fatty acid binding protein 4 is a target of VEGF and a regulator of cell proliferation in endothelial cells. *The FASEB Journal, 23*, 3865–3873.

Ferrara, N., Frantz, G., Lecouter, J., Dillard-Telm, L., Pham, T., Draksharapu, A., Giordano, T., & Peale, F. (2003a). Differential expression of the angiogenic factor genes vascular endothelial growth factor (VEGF) and endocrine gland-derived VEGF in normal and polycystic human ovaries. *American Journal of Pathology, 162*, 1881–1893.

Ferrara, N., Gerber, H. P., & Lecouter, J. (2003b). The biology of VEGF and its receptors. *Nature Medicine, 9*, 669–676.

Folkman, J., & Klagsbrun, M. (1987). Vascular physiology. A family of angiogenic peptides. *Nature, 329*, 671–672.

Gealekman, O., Burkart, A., Chouinard, M., Nicoloro, S. M., Straubhaar, J., & Corvera, S. (2008). Enhanced angiogenesis in obesity and in response to PPARgamma activators through adipocyte VEGF and ANGPTL4 production. *American Journal of Physiology. Endocrinology and Metabolism, 295*, E1056–E1064.

Hu, Y., Eastabrook, G., Tan, R., Maccalman, C. D., Dutz, J. P., & von Dadelszen, P. (2010). Decidual NK cell-derived conditioned medium enhances capillary tube and network organization in an extravillous cytotrophoblast cell line. *Placenta, 31*, 213–21.

Johnsen, G. M., Basak, S., Weedon-Fekjaer, M. S., Staff, A. C., & Duttaroy, A. K. (2011). Docosahexaenoic acid stimulates tube formation in first trimester trophoblast cells, HTR8/SVneo. *Placenta, 32*, 626–632.

Lyall, F., Bulmer, J. N., Duffie, E., Cousins, F., Theriault, A., & Robson, S. C. (2001). Human trophoblast invasion and spiral artery transformation: The role of PECAM-1 in normal pregnancy, preeclampsia, and fetal growth restriction. *American Journal of Pathology, 158*, 1713–1721.

Massaro, M., Scoditti, E., Carluccio, M. A., Campana, M. C., & de Caterina, R. (2010). Omega-3 fatty acids, inflammation and angiogenesis: Basic mechanisms behind the cardioprotective effects of fish and fish oils. *Cellular and Molecular Biology, 56*, 59–82.

Moon, E. J., Lee, Y. M., & Kim, K. W. (2003). Anti-angiogenic activity of conjugated linoleic acid on basic fibroblast growth factor-induced angiogenesis. *Oncology Reports, 10*, 617–621.

Mudgett, J. S., Ding, J., Guh-Siesel, L., Chartrain, N. A., Yang, L., Gopal, S., & Shen, M. M. (2000). Essential role for p38alpha mitogen-activated protein kinase in placental angiogenesis.

Proceedings of the National Academy of Sciences of the United States of America, 97, 10454–10459.

Murota, S., Kanayasu, T., Nakano-Hayashi, J., & Morita, I. (1991). Involvement of eicosanoids in angiogenesis. *Advances in Prostaglandin, Thromboxane, and Leukotriene Research, 21B*, 623–626.

Myatt, L. (2002). Role of placenta in preeclampsia. *Endocrine, 19*, 103–111.

Reynolds, L. P., & Redmer, D. A. (2001). Angiogenesis in the placenta. *Biology of Reproduction, 64*, 1033–1040.

Shibuya, M. (2006). Vascular endothelial growth factor receptor-1 (VEGFR-1/Flt-1): A dual regulator for angiogenesis. *Angiogenesis, 9*, 225–230. discussion 231.

Spencer, L., Mann, C., Metcalfe, M., Webb, M., Pollard, C., Spencer, D., Berry, D., Steward, W., & Dennison, A. (2009). The effect of omega-3 FAs on tumour angiogenesis and their therapeutic potential. *European Journal of Cancer, 45*, 2077–2086.

Taki, A., Abe, M., Komaki, M., Oku, K., Iseki, S., Mizutani, S., & Morita, I. (2012). Expression of angiogenesis-related factors and inflammatory cytokines in placenta and umbilical vessels in pregnancies with preeclampsia and chorioamnionitis/funisitis. *Congenital Anomalies (Kyoto), 52*, 97–103.

Tobin, K. A., Harsem, N. K., Dalen, K. T., Staff, A. C., Nebb, H. I., & Duttaroy, A. K. (2006). Regulation of ADRP expression by long-chain polyunsaturated fatty acids in BeWo cells, a human placental choriocarcinoma cell line. *Journal of Lipid Research, 47*, 815–823.

Valdes, G., & Corthorn, J. (2011). Review: The angiogenic and vasodilatory utero-placental network. *Placenta, 32*(Suppl 2), S170–S175.

Wang, K., Jiang, Y. Z., Chen, D. B., & Zheng, J. (2009). Hypoxia enhances FGF2- and VEGF-stimulated human placental artery endothelial cell proliferation: Roles of MEK1/2/ERK1/2 and PI3K/AKT1 pathways. *Placenta, 30*, 1045–1051.

Zhang, Y., & Daaka, Y. (2011). PGE2 promotes angiogenesis through EP4 and PKA Cgamma pathway. *Blood, 118*, 5355–5364.

Zhong, Y., Tuuli, M., & Odibo, A. O. (2010). First-trimester assessment of placenta function and the prediction of preeclampsia and intrauterine growth restriction. *Prenatal Diagnosis, 30*, 293–308.

Zhu, R., Huang, Y. H., Tao, Y., Wang, S. C., Sun, C., Piao, H. L., Wang, X. Q., Du, M. R., & Li, D. J. (2013). Hyaluronan up-regulates growth and invasion of trophoblasts in an autocrine manner via PI3K/AKT and MAPK/ERK1/2 pathways in early human pregnancy. *Placenta, 34*, 784–791.

Chapter 3
Glucose and Amino Acid and Their Roles in Placentation

Contents

3.1 Introduction

The "Fetal origins hypothesis" states that individual born small due to maternal malnutrition is predisposed to adult disease. The hypothesis is skewed more towards malnourished mothers and does not specify to include placental factor. In contrast, poor placentation mainly due to inadequate vascular adaptation at the utero-placental interface is far more reasonable cause of reduced fetal growth in adequately nourished populations (Henriksen and Clausen 2002). Now, substantial evidences exist that suboptimal maternal nutrition can regulate newborn in both the cases. Immediate impact of optimum maternal diet or its deviation on placental growth and fetal birth weight would be the key to understand their beneficial effect on placentation process. The availability and supply of nutrients to the developing fetus depend on maternal nutritional status which in turn depends on their nutrient stores, dietary intake, and obligatory requirements during pregnancy (Ramakrishnan et al. 2012). During development, there are critical periods when system and organs undergo maturation; those stages are susceptible to be programmed. The pregnancy-specific stages that span from the preconception to the birth are subjected to be influenced by macro and micronutrients that affect both maternal health and fetus development subsequently. Data so far highlighted the

A.K. Duttaroy, S. Basak, *Early Nutrition and Lifestyle Factors*,
DOI 10.1007/978-3-319-38804-5_3

importance of nutrition during pregnancy typically focusing on the second and/or the third trimester by which time major organogenesis would have been completed. However, nutritional impact just before conception and/or during first trimester, when women are typically unaware of their pregnancy status, is scanty. It is likely that preconception, conception, implantation, organogenesis, and placentation will be influenced by status of the maternal periconceptional nutrition. These effects may lead to regulate overall health of childbearing women, their reproductive potential, and birth outcome.

3.2 Role of Early Nutrients: An Overview

The nutritional status of women is the key for successful pregnancy. In particular, diet during the first trimester may be more important to support placental formation and the development and differentiation of various organs. Moreover, preconceptional nutrition is crucial for an optimal onset and reproductive potential of women. Principal nutrients in the diet that are required in a large quantity to meet daily energy and substrate requirement are termed as macronutrients. Glucose is the primary energy source for the fetus metabolism at basal level and for protein synthesis and source of glycogen storage and fat. Amino acids are the building blocks for protein synthesis, and growth provides oxidative substrate in the absence of glucose. Fatty acids of omega-3 ($n-3$) and omega-6 ($n-6$) serve as energy store as well as used for membrane formation (specially important for brain and retina during the last trimester of pregnancy) and modulate biological membranes by influencing properties of the membrane fluidity, cell–cell interaction, and intracellular transporting and signaling function of their eicosanoid derivatives. Embryo stores intracellular lipids in the form of lipid droplets (Tsujii et al. 2001). Vitamins and minerals are required in a small quantity often termed as micronutrients. These are primarily coenzymes and cofactors of many enzymatic pathways. In spite of their low requirements, their effects are far more qualitative for the fetus outcome. Folic acid is increasingly required for rapid growth and cell proliferation during initial phases of pregnancy. Its deficiency leads to congenital malformations. In developing countries, diets are generally low in animal products and consequently in vitamin B_{12} content. An insufficient supply may cause reduced fetal growth. Vitamin D is thought to perform an immunomodulatory role to complete the trophoblast invasion of the uterus without triggering a maternal immune response. Vitamin A deficiency and excess have profound effects on the development of the vertebrate embryo. Vitamin B_6 acts as coenzyme in protein metabolism and development of the central nervous system. Vitamin C (ascorbic acid) and vitamin E (alpha-tocopherol) are powerful antioxidants that have been used for the prevention and treatment of preeclampsia. Maternal iron (Fe) deficiency has a direct impact on neonatal Fe stores and birth weight, and its deficiency may cause cognitive and behavioral problems in childhood. Calcium (Ca) deficiency is associated with preeclampsia and intrauterine growth restriction. Magnesium (Mg) deficiency may cause hematological and teratogenic damage. Selenium (Se) is an antioxidant that supports humoral and cell-mediated immunity. Low Se status is associated with recurrent abortion, preeclampsia, and IUGR. Microenvironment also regulates the success of placentation. Oxygen is a major regulator of trophoblast invasion. Thus, they are high in the first trimester when oxygen tension in the placental bed is low and trophoblast invasion occurs and then fall at 10–12 weeks of gestation when blood flow to the intervillous space is established and oxygen tension rises. In vitro, low oxygen tension (hypoxia) promotes trophoblast differentiation down the EVT pathway. The exchange of nutrients between placenta and fetus involves three major mechanisms: (1) direct placental transfer of nutrients from the mother to the fetal plasma, (2) placental metabolism and consumption of nutrients, and (3) placental metabolism of nutrient substrates to alternate substrate forms. Placenta performed dual role in determining the outcome of the fetus. In one hand, placenta protects fetus from toxic xenobiotic compounds by forming an immunological barrier, on the other hand, it forms nutrient transporting corridor to establish the fetal–maternal interactions that is critical for successful human pregnancy. In achieving successful placentation, trophoblast invasion into uterus is the key event that has been compromised in many pathological pregnancies. All these nutrients are indispensable for the development of the placenta. In this chapter, role of glucose and amino acids (key macronutrients) on placentation will be discussed, while other essential macronutrients, e.g., fatty acids and their synthesis, transport, deficiency, and gene defects, will be dealt in separate chapters in the later part of this book.

3.3 Glucose Transport During Placentation

Efficiency of the placental transport capacity depends on several factors that include surface areas for exchange, thickness of the internal membrane, and density of the transporter proteins integrated into the trophoblast membrane. The process of uterine preparation for embryo receptivity is classically known as decidualization where differentiation of endometrial stromal cells (ESCs) into decidual cells involves rapid participation of glucose metabolism. Glucose is the primary energy substrate for the oxidative metabolism of the fetus growth and development. Studies have implicated the role of steroid hormones, E2 and P4, in the regulation of glucose metabolism by regulating glucose transporter (GLUT) expression. The GLUTs are responsible for glucose uptake by cells and, thus, play a critical role in normal glucose utilization. Although, glucose transport is concentration dependent across the placenta, however, it is facilitated by a family of glucose carriers known as glucose transporters (gene symbol SLC2A, protein symbol GLUT). So far, 14 isoforms of the membrane-spanning glucose transporter are reported, in which GLUT1 and GLUT3 are the key members expressed in the human placental trophoblast (Illsley 2000). Key glucose transporters of the first trimester placenta are listed in Table 3.1. GLUT1 and GLUT3 are known to be expressed in human

Table 3.1 Key glucose transporters of the first trimester placenta

Type	Function	References
GLUT 1	Increased GLUT1 transporter activity improves decidualization in uterine stromal cells	Frolova et al. (2009)
	Major form of glucose transporter in early pregnancy; increased uptake of glucose in the first trimester placenta and positive immunostaining with GLUT1 and GLUT3	von Wolff et al. (2003)
	Maternal glucose abnormality linked with placental GLUT1 expression	Gordon et al. (1995)
	Activation of decidual leucocytes during implantation and improving immunological tolerance	Korgun et al. (2005)
GLUT 3	Sustained activity of GLUT3 in early trimester suggests for possible role in energy supply of early placentation	Korgun et al. (2005)
GLUT 12	GLUT12 mRNA expressed in first trimester placenta and absent in term placenta	Gude et al. (2003)
	Maternal insulin regulated glucose uptake by GLUT12 in the first trimester trophoblast	Gude et al. (2003)

endometrium, and GLUT1 expression increased upon decidualization of the uterine stromal cells in vitro. GLUT1 transporter activity is common in ESC and positively regulates the process of decidualization (Frolova et al. 2009). Therefore, maternal conditions associated with altered glucose utilization may impair implantation at the level of ESCs (Frolova and Moley 2011). Developing fetus obtains most of its energy from the metabolism of glucose derived from the mother. Therefore, transport of glucose from the mother to fetus is a key event that mediates via placenta. Glucose uptake by the placental trophoblast is regulated by many factors that may affect several processes such as cell proliferation, viability, and migration of the trophoblast cells. Glucose uptake of the first trimester HTR8/SVneo trophoblast cells was inhibited by xanthohumol that impairs cell viability and proliferation of these cells. Insufficient glucose uptake during early trimester leads to adverse effects on cell growth and differentiation process that may cause inappropriate growth and development of the trophoblast (Correia-Branco et al. 2015). Uptake of ^{3}H-deoxy-D-glucose is found to be time-dependent, saturable, and insulin-insensitive process in the HTR8/SVneo cells (Correia-Branco et al. 2015). Adequate glucose metabolism is required to provide nutritional support of the endometrium for successful decidualization. Glucose uptake is facilitated by glucose transporter molecules during implantation and creates optimum microenvironment for receptive milieu. In human endometrium, only GLUT1 and GLUT3 are expressed, whereas other insulin-dependent glucose transporters such as GLUT2, GLUT4, and GLUT8 are not detected. Inhibition of glucose transporters by cytochalasin B decreased the uptake of glucose in stromal decidualization (von Wolff et al. 2003). The uptake of glucose in the placenta is facilitated by GLUT1 and GLUT3 proteins during first trimester of pregnancy. It is evidenced by positive immunostaining of GLUT1 and GLUT3 with the tissue sections of first trimester human villous placenta. The negative immunostaining for GLUT4 protein in the human feto-maternal interface suggests that GLUT4 may not be expressed in the first trimester placenta. Data from several studies concluded that GLUT1 is the major glucose transporter in the placenta, especially in early trimester of the human pregnancy (Hahn et al. 1998). GLUT1 protein is positively stained with the syncytial microvillous that is considered as the primary source of the glucose for the fetus. GLUT1 expression is detected in cytotrophoblast cells, and fetal endothelial cells of the first trimester placental villi suggest their requirement to perform different function associations with implantation (Hahn et al. 2000; Illsley et al. 1998). GLUT1 accounts for major binding sites of cytochalasin B in the placental membranes (Barros et al. 1995). GLUT3 is localized in the endothelial cells lining of the fetal capillaries and thereby maintaining glucose homeostasis between these cells and fetus. The presence of GLUT3 protein is not very prominent in the term human placenta (Janzen et al. 2013; Jansson et al. 1993). Since, GLUT3 protein is expressed in the invasive choriocarcinoma cell lines, it is presumed that GLUT3 may be required in the metabolically active and invading first trimester trophoblast cells (Illsley et al. 1998). Glucose provides energy support to maternal leucocytes. Reports suggest that sustained activities of GLUT3 are probably required in providing energy for the proliferation and invasion of the extravillous trophoblast during implantation process (Korgun et al. 2005). The role of GLUT1 on fetal angiogenesis is predicated, since this protein is localized in the capillary tube formation of microvascular connecting tubes formed by one or several angiogenic cells (Demir et al. 2004). GLUT1 and GLUT3, are shown to express in CD45 and CD68 positive cells, suggest their role in the activation of the decidual leukocytes during implantation and thereby improving immunological tolerance (Korgun et al. 2005). Another glucose transporter, GLUT12, is reported to be expressed in the first trimester placenta. Structurally, GLUT12 amino acids exhibit 40 % homology with GLUT10 and 29 % with GLUT4. Expression of GLUT12 mRNA is observed in placental tissue and trophoblast cells (Gude et al. 2003). Interestingly, GLUT12 is predominantly expressed in the syncytiotrophoblast as well as in extravillous trophoblast cells of the first trimester placental tissues of 10–12 weeks of gestation, but not in the term syncytiotrophoblast which implies its specific role in the first trimester of the pregnancy. GLUT12 may play a role in the uptake of glucose by human placental cells and may be an important additional pathway for the regulation of glucose handling by the human placenta. GLUT12 is thought be an insulin-sensitive glucose transport system in placenta. In contrast to term syncytiotrophoblast, the specific presence of GLUT12 together with insulin receptor in first trimester syncytiotrophoblast suggests that maternal insulin may regulate the glucose uptake of the placenta via GLUT12 during early pregnancy.

3.4 Regulation of Glucose Transporters in Feto-Placental Unit

Activities of the glucose transporters are highly regulated by hyperglycemia. In diabetic mice, glucose concentration negatively regulates the expression of glucose transporter, GLUT1 (Ogura et al. 1999). GLUT1 activity is regulated by several factors in the feto-placental unit (Table 3.2). Glucose and insulin combination modulates the mRNA expression of glucose transporter and glucose uptake in trophoblast isolated from first-trimester chorionic villi. Maternal glucose abnormalities and impaired fetal development during the first trimester have been linked with placental GLUT1 expression (Gordon et al. 1995). Direct effects of insulin growth factor (IGF-1) on trophoblast cells showed an increase in the expression of cellular GLUT1 protein in BeWo choriocarcinoma cells. Irrespective of its action on either via microvillous or basal membrane receptors, trans-epithelial glucose transport as well as GLUT1 expression is increased in these cells. Insulin growth factor possibly controls the nutrient supply in regulating the fetal growth by modulating the activities of the GLUT (Baumann et al. 2014). Maternal hyperglycemia downregulates the GLUT1 transport system of term placental trophoblast by redistribution of the localization of the GLUT1 protein (Hahn et al. 2000). The GLUT1 expression is reduced significantly when exposed to 25 mmol/L glucose in vitro due to internalization of the plasma membrane of GLUT1 protein. Excess glucose (25 mmol/L) may lead to translocation of GLUT1 protein from the trophoblast surface to intracellular sites and thereby self-regulates hexose transport by altering GLUT1 partition between the plasma membrane and intracellular sites in term placental trophoblast cells. Exogenous D-glucose does not alter the activity and the expression of GLUT1 in other trophoblast cell lines such as JAR and JEG-3.

Table 3.2 Regulation of GLUT1 activity in the feto-placental unit

Function	References
Abundant in maternal and fetal facing side but undetected in fetal capillary	Barros et al. (1995)
Insulin stimulates GLUT1 expression in vitro	Baumann et al. (2014)
Increased GLUT1 expression in placenta positively correlate with birth weight	Acosta et al. (2015)
GLUT1 expression and glucose uptake unaffected in IUGR	Jansson et al. (2002)
Hypoxia stimulates GLUT1 expression and increased placental transport of glucose	Baumann et al. (2007), Hayashi et al. (2004)
GLUT1 transporter protein expression decreased in severe hypoxia	Zamudio et al. (2010)
GLUT1 expression and glucose uptake increased by resistin in trophoblast cell	Di Simone et al. (2009)
Glucose concentration negatively regulates expression of GLUT1 in diabetic mice	Ogura et al. (1999)
Maternal hyperglycemia downregulates GLUT1 transport of the placenta	Hahn et al. (2000)

Neither glucose transport activity in JEG-3 cells is affected by extracellular glucose treatment (Illsley et al. 1998). Placental glucose transporter activity is correlated with birth weight of the fetus. Basal plasma membrane GLUT1 expression of the placentas ($n = 33$) is positively correlated with birth weight. Since maternal fasting glucose and placental glucose transport capacity are not increased in pregnant obese women, therefore, it is assumed that increased placental size promoted glucose delivery to the fetus (Acosta et al. 2015). Reduced level of fetal plasma concentrations of amino acids and glucose is associated with intrauterine growth restriction (IUGR). The activity of the amino acid transporter system A in the syncytiotrophoblast microvillous membrane (MVM) is reduced in IUGR. However, GLUT1 expression and D-glucose uptake are not affected by IUGR (Jansson et al. 1993, 2002; Janzen et al. 2013). The hypoxia seems to regulate the expression of GLUT proteins. The effect of hypoxia on glucose transport and metabolism is studied in human trophoblast cells derived from term placenta. Trophoblast exposed to hypoxia (PO_2, 12–14 mmHg) conditions leads to the changes in the expression of glucose transporters (GLUT1 and GLUT3) compared to normoxia (PO_2, 120–130 mmHg). Hypoxic trophoblast cells are loaded with increased numbers of mitochondria, lysosomes, and vacuoles. It seems that trophoblast in culture survives in extreme hypoxia, but manifests striking changes in morphology and in glucose metabolism and transport (Esterman et al. 1997). Many of the in vitro studies demonstrated that reduction in the oxygen tension stimulates the expression of GLUT1 and GLUT3 and thereby increasing trans-epithelial glucose transport which is mediated by the action of hypoxia-inducible transcription factor-1α (HIF-1α) (Baumann et al. 2007; Hayashi et al. 2004; Esterman et al. 1997). Under hypoxia, placental transport of glucose potentially increases the concentration of glucose as an alternate metabolic fuel to both placenta and the fetus as a part of the adaptive response, since gluconeogenesis pathway does not operate efficiently in placenta. Paradoxically, human fetus affected by IUGR demonstrates hypoglycemia; the effect is correlated with severity of growth restriction and decreased blood flow (Economides et al. 1990). Altered feto-placental glucose metabolism is responsible for reduced fetal growth at high altitude where reduced oxygenation prevails. Under this condition, umbilical venous and arterial glucose concentrations are markedly decreased, resulting in lower glucose delivery to the fetus. In addition to reduced glucose levels, fetal lactate levels are increased, insulin concentrations are decreased, and the expression of GLUT1 in the placental basal membrane is reduced. Therefore, it is suggested that placenta-mediated reduction in glucose transport is the key factor for reduced fetal growth under condition of severe hypoxia (Zamudio et al. 2010). Glucose uptake and transporter activities are modulated by adipocytokines in human trophoblast cells. Resistin increases placental glucose uptake and expression of GLUT1 by activation of ERK-1 and ERK-2 pathways in trophoblast cells (Di Simone et al. 2009).

3.5 Glucose and Early Trophoblast Development

Extravillous trophoblast (EVT) cells may behave in a similar way as with endothelial cells with respect to angiogenesis processes during endovascular differentiation that involves remodeling of spiral arteries in order to improve blood flow towards the intravillous space. Endovascular differentiation involves coordinated action of key growth factors VEGF and integrin signaling in the establishment and maintenance of successful placentation (Fukushima et al. 2005). The invasion and migration of trophoblast cells are critical for normal placentation that ultimately leads to a successful pregnancy. The whole process of placentation is a well-coordinated network of several factors at the fetal–maternal surface. Inadequate uterine invasion by trophoblast cells may lead to poor perfusion of the placenta or complications such as preeclampsia. Glycogen metabolism is abnormal in trophoblast-related pregnancy disease such as preeclampsia. Many studies investigated energy metabolism and glycolysis of term placental trophoblast in order to understand their role in differentiation of cytotrophoblast and syncytiotrophoblast of the placenta. During increased energy demand, cytotrophoblast cells derive energy from both glycolytic and non-glycolytic pathways, while syncytiotrophoblast meets energy requirement from aerobic metabolism (Bax and Bloxam 1997). Excess glucose in the form of hyperglycemia disrupts the invasive profile of cytotrophoblast by decreasing the urokinase plasminogen activator (uPA) activity, downregulating VEGF and PIGF, and upregulating sEng, sFlt-1, and interleukin-6. These data suggest that hyperglycemia may mediate their effects on cytotrophoblast invasion by activation of stress signaling pathway (Cawyer et al. 2014). Effect of glucose on the first trimester trophoblast cells, HTR8/SVneo, showed a reduction in invasiveness of the trophoblast by lower activity of urokinase plasminogen activator (Belkacemi et al. 2005). Early intrauterine effects of high glucose as happened during diabetes cause a reduction in the growth of the placenta. Growth and proliferation on the cellular models of trophoblast (BeWo, JAR and JEG-3 choriocarcinoma cells) are reduced in the presence of 25 mmol/L D-glucose as compared to osmotic control. The mechanism that leads to growth arrest of the trophoblast cells in vitro may relate with the alteration in the mitochondrial activities and differentiation of the cells. Authors suggest that hyperglycemia-induced placental growth restriction could be manifested in the first trimester of a poorly controlled diabetic pregnancy (Weiss et al. 2001). Impacts of maternal hyperglycemia on early placentation processes that involves tube formation, cellular growth and proliferation, and metabolic activities of the first trimester trophoblast cells were mimicked in vitro. Designing appropriate dose for glucose in cell culture that mimics hyperglycemia is the limitation of all the in vitro studies with glucose treatment, since majority of the cell culture media contains glucose more than 10 mM. Study on HTR8/SVneo cells showed increased tube formation and viability of the cells when cultured with 25 mM glucose without any cytotoxicity. Moreover, glucose (25 mM) significantly increased mRNA and protein level of matrix metalloproteinase-9 (Basak et al. 2015). The effect of hyperglycemia-induced alterations of the cell surface of proteoglycans and the ECM remodeling on the expressions of angiogenesis-related cytokines and growth factors is observed in 3A-Sub-E trophoblastic cells. All these data suggest that altered placental structure and maternal–fetal complication during development may be the initial response to glucose challenge of the first trimester trophoblast cells (Chang and Vivian Yang 2013).

3.6 Amino Acid Transport and Its Regulation in Early Pregnancy

Amino acids are building blocks for the protein synthesis required for the muscle and lean body mass development of the fetus and provides energy during development of the fetus. Essential, nonessential, and conditionally essential amino acids are classified based on the requirement of the protein synthesis. An additional group, known as functional amino acids that regulate various pathways of growth and survival, development, and reproduction of humans, includes nutrients such as arginine, cysteine (Cys), glutamine (Gln), leucine (Leu), proline (pro), and tryptophan (Trp). Distribution of amino acids inside the first trimester gestational sac is measured in placental villi of early gestation (7–11 weeks) and in samples of coelomic and amniotic fluid. The highest levels of amino acids are found, taurine, glutamic acid, glycine, and alanine, in the first trimester placental villi. Significant positive association of key amino acids between maternal serum and coelomic fluid indicated that transport and accumulation of amino acid take place as early as 7 weeks of gestation (Jauniaux et al. 1998). Maternal transport of amino acids across the placenta undergoes via active concentration gradient system using Na^+-dependent cotransporters where energy from Na^+ gradient is established by Na^+/K^+ ATPase (Bazer et al. 2015). System A amino acid transporter is a Na^+-dependent cotransporter that transport neutral amino acids such as alanine, serine, and glycine across the placental membrane. Methylaminoisobutyric acid (MeAIB), a non-metabolizable substrate specific for system A transporter, is used to estimate the activity of this transporter in a number of human studies. There are three subtypes of system A transporter that are expressed in human placental trophoblast cells (SNAT1, 2, and 4). Placental system A transporter activity increases with gestational age. SNAT4-mediated system A activity is higher in first trimester than at term, whereas SNAT1 and/or SNAT2 system A activity is increased later in gestation (Desforges et al. 2010). Placental system A transporter activity is inversely associated with placental weight. Thus, smaller the placentas greater is the upregulation of system A transport, per gram of placenta (Dilworth and Sibley 2013). Limited data are available on how functional amino acids and its transporters regulate pathways that are critical for the establishment of feto-placental unit during early human pregnancy. Small size placentas that are exposed to maternal undernutrition exhibited altered structure with enlarged labyrinth exchange zone at the expense of endocrine junction zone. These structural changes are linked with increased transport of non-metabolic amino acid analogues per weight of the placenta, and expression of the genes encodes System A transporters. Inadequate nutrient supply or small size of placenta, when becomes limiting factor for normal fetal growth, placental plasticity works extra to increase its transport capacity (Burton and Fowden 2012). Maternal protein restriction in the rat affects placental insulin and mammalian target of mTOR signaling, with considerable downregulation of placental microvillous plasma membranes system A and L amino acid transporter activity. It is suggested that derangement of maternal endocrine and metabolic control of placental transport reduces fetal growth in protein-restricted dams (Rosario et al. 2011). Both glucose transporter 1 (GLUT1) and amino acid transporter system A activity of the term placenta are increased in pregnancies complicated by diabetes. In contrast, placental amino acid transporter system A activity is decreased in IUGR. These transporters' activities are not influenced when exposed briefly with leptin, insulin, and glucose in the first trimester villous fragments (Ericsson et al. 2005). Limited nutrient availability to the fetus and insufficient weight gain during pregnancy are two characteristics of the pathological pregnancy such as IUGR and PE. Adaptive response of the placenta is often associated with increased expression of amino acid transporters. Interestingly, in case where overall availability of the nutrients is low such as PE or IUGR, an overall increase in the level of system L and mTOR protein expression follows to syncytiotrophoblast microvillous membrane. A basis of the association between amino acid transporters expression and pre-pregnancy BMI may be exploited further (Aiko et al. 2014). Trophoblast amino acid transport and signaling is regulated by long chain fatty acids. Combination of docosahexenoic acid (DHA) with oleic acid (OA) increased amino acid transport and phosphorylation of ERK and mTOR in primary human trophoblasts (Lager et al. 2014). This study further suggests the importance of long chain fatty acids for the placental function in the regulation of the amino acid transporters of the developing fetus. The amino acid transport systems such as A and L preferably transfer essential amino acids. Activity of the placental amino acid transport systems A and L is regulated in the presence of hypoxia in the villous placental trophoblst. Hypoxia (2 % O_2) decreased system A-mediated methyl aminoisobutyric acid (MeAIB) uptake (system A) and increased H^3-leucin (system L) uptake compared to standard culture conditions (21 % O_2) in the primary villous fragments isolated from term placentas (Kleppa et al. 2014). Nitric oxide (NO) is an essential component to develop a healthy endothelium and promotes endovascular invasion by the trophoblast during early placentation. L-arginine is the substrate for the enzyme NO synthase, and asymmetric dimethylarginine (ADMA) is an endogenous inhibitor of this enzyme. The ratio of L-arginine to ADMA is associated with altered NOS activity. Trophoblast invades maternal spiral artery across the uterine wall and produces NO which acts on artery wall to create low resistance high flow of the bloods across the feto-placental unit. Maternal plasma level of L-arginine at first trimester gestation is decreased in pregnancy where trophoblast invasion is inadequate and subsequently develops early PE (<34 weeks) (Khalil et al. 2015). High level of ADMA in PE is

Table 3.3 Regulation of amino acid transporter and protein activity in the first trimester placenta

Parameters	Function	References
System A transporter	Highest level in neutral amino acids of the first trimester amniotic fluid (7–11 weeks)	Jauniaux et al. (1998)
System A transporter subtype	Higher activity of SNAT4-mediated system A transporter in early gestation compared to SNAT-1 and SNAT-2	Desforges et al. (2010)
System A transporter	Increase in placental system A transporter results in the decrease in placenta weight	Dilworth and Sibley (2013)
L-Arginine level	Decrease in L-arginine level in first trimester maternal plasma that later develops PE	Khalil et al. (2015)
Protein deficiency	Gestational protein deficiency impairs follicle development in fetus	Almeida et al. (2012)
Protein-energy deficiency	Maternal malnutrition impairs vascular development in rat fetus	Ferreira et al. (2010)
Protein deficiency	Impaired angiogenesis of the animal placenta	Lagarde et al. (2014)
Adolescent pregnancy	Reduced uterine blood flow and impaired placental vascular development in ewes	Luther et al. (2007)
Protein deprivation	Alteration in the placental epigenetic pathway in early gestation of mice	Gheorghe et al. (2009)

assumed to be responsible for compromised NO synthesis that contributes endothelial dysfunction and impaired placentation (Alpoim et al. 2013). Activity of amino acid transporters is regulated by different factors in the first trimester placenta (Table 3.3). Maternal obesity is linked with fetal overgrowth and macrosemia. However, actual role of the placenta in linking this process is largely unknown. Specific upregulation of placental system A transporter activity and activation of insulin/IGF-1 and mammalian target of rapamycin (mTOR) signaling are observed in placenta samples of BMI matched ($n = 23$) obese pregnant women (Jansson et al. 2013). Maternal vitamin D level during pregnancy and placental amino acid transporter gene expression are associated with development of the offspring in terms of body composition and bone metabolism, since many of the amino acid transporter genes have vitamin D response elements in their promoters. A study with human placenta samples ($n = 102$) demonstrated that expression of amino acid transporter genes are closely associated with the maternal vitamin D level (Cleal et al. 2015).

3.7 Protein Malnutrition and Placentation

Pre and postnatal protein deficiency often leads to decreased fetal intrauterine development and postnatal growth, which is widespread in many developing countries. Gestational protein deficiency affects ovaries of female rats' offspring. The protein-deficient female offspring had a smaller ovarian area, greater numbers of primordial follicles and developing follicles per square millimeters of ovarian cortex, and no corpora lutea (Almeida et al. 2012). Thus, protein-deficient female offspring showed gross histomorphological changes of follicle development. Maternal malnutrition during lactation programmed the follicular development by reducing mRNA expression of VEGF and FGF with a reduction of vasculature that possibly leads to supply less folliculotrophic substances to these protein-energy deficient rats (Ferreira et al. 2010). Adolescent pregnancy is one form of energy-deficient pregnancy malnutrition which is manifested with suboptimal dietary intake of nutrients associated with inadequate gestational weight gains. Therefore, human adolescent mother constitutes a proxy indicator of maternal undernutrition. In such scenario, fetal growth restriction is reported due to reduced availability of the nutrient from maternal circulation. In addition to that, alternation in maternal placental vascular development further contributes to the overall reduction of the nutrients availability for the fetus. Placental vascularization determines blood flow (both volume and rate) to the uterus which is also a determinant for the fetal growth. Reduced capillary area density of the vascular bed implied a compromised uterine blood flow with reduced capacity in the undernourished ewes (Luther et al. 2007). In human, increased uterine vascular resistance and reduced uterine blood flow indicates altered placental vascular development. Maternal undernutrition during pregnancy impairs placental angiogenesis and vascular development in humans. The effects of protein malnutrition in utero on vascular growth of the fetal lung and placenta led to higher glycogen levels, but reduced number of alveoli as confirmed by the measurement of radial alveolar counts. The protein-deficient animal fetus had significantly reduced levels of angiogenic growth factors such as VEGF in lung and decreased expression of VEGF in the placenta (Lagarde et al. 2014). Protein deprivation in early pregnancy negatively regulates expression of gene encoding cellular growth and metabolism, while genes involved in epigenetic pathways are upregulated. Thus, effect of protein malnutrition is not only restricted to placental alteration in the morphology but also induces alteration in genome imprinting of the placenta (Gheorghe et al. 2009).

3.8 Summary

Glucose is the primary energy substrate for feto-placental growth and development. Developing fetus obtains most of its energy from the metabolism of glucose derived from the mother. Adequate glucose metabolism is required to provide nutritional support of the endometrium for successful decidualization. Glucose is taken up via GLUT1 and GLUT3 transporters from the maternal circulation. These are not only required for placental activities and function but also needed for the successful implantation. Placental glucose transporter activity is correlated with birth weight of the fetus. Maternal hyperglycemia downregulates the activities of the GLUT1 transporter by redistribution of GLUT1 protein accessibility. Exogenous glucose showed reduced invasiveness of the trophoblast in vitro. Unlike glucose, amino acids are actively transported and accumulated as early as 7 weeks of gestation.

SNAT4-mediated system A activity is found higher in first trimester pregnancy than at term, whereas SNAT1 and/or SNAT2 system A activity is increased later in gestation. Placental system A transporter activity is inversely associated with placental weight. A positive association is observed between amino acid transporters expression and pre-pregnancy BMI. Maternal plasma level of L-arginine is decreased at first trimester gestation where trophoblast invasion is inadequate and subsequently develops PE. Macronutrients such as glucose, amino acids, and their transporter systems are expressed in the first trimester placental trophoblast. Regulation of these transporters related with the development of normal and altered placentation warrants further studies in first trimester trophoblast involving early pregnancy.

References

Acosta, O., Ramirez, V. I., Lager, S., Gaccioli, F., Dudley, D. J., Powell, T. L., & Jansson, T. (2015). Increased glucose and placental GLUT-1 in large infants of obese nondiabetic mothers. *American Journal of Obstetrics and Gynecology, 212*, 227.e1–227.e7.

Aiko, Y., Askew, D. J., Aramaki, S., Myoga, M., Tomonaga, C., Hachisuga, T., Suga, R., Kawamoto, T., Tsuji, M., & Shibata, E. (2014). Differential levels of amino acid transporters System L and ASCT2, and the mTOR protein in placenta of preeclampsia and IUGR. *BMC Pregnancy and Childbirth, 14*, 181.

Almeida, F. R., Silva, G. A., Fiuza, A. T., Chianca, D. A., Jr., Ferreira, A. J., & Chiarini-Garcia, H. (2012). Gestational and postnatal protein deficiency affects postnatal development and histomorphometry of liver, kidneys, and ovaries of female rats' offspring. *Applied Physiology, Nutrition, and Metabolism, 37*, 293–300.

Alpoim, P. N., Godoi, L. C., Freitas, L. G., Gomes, K. B., & Dusse, L. M. (2013). Assessment of L-arginine asymmetric 1 dimethyl (ADMA) in early-onset and late-onset (severe) preeclampsia. *Nitric Oxide, 33*, 81–82.

Barros, L. F., Yudilevich, D. L., Jarvis, S. M., Beaumont, N., & Baldwin, S. A. (1995). Quantitation and immunolocalization of glucose transporters in the human placenta. *Placenta, 16*, 623–633.

Basak, S., Das, M. K., Srinivas, V., & Duttaroy, A. K. (2015). The interplay between glucose and fatty acids on tube formation and fatty acid uptake in the first trimester trophoblast cells, HTR8/SVneo. *Molecular and Cellular Biochemistry, 401*, 11–19.

Baumann, M. U., Schneider, H., Malek, A., Palta, V., Surbek, D. V., Sager, R., Zamudio, S., & Illsley, N. P. (2014). Regulation of human trophoblast GLUT1 glucose transporter by insulin-like growth factor I (IGF-I). *PLoS One, 9*, e106037.

Baumann, M. U., Zamudio, S., & Illsley, N. P. (2007). Hypoxic upregulation of glucose transporters in BeWo choriocarcinoma cells is mediated by hypoxia-inducible factor-1. *American Journal of Physiology – Cellular Physiology, 293*, C477–C485.

Bax, B. E., & Bloxam, D. L. (1997). Energy metabolism and glycolysis in human placental trophoblast cells during differentiation. *Biochimica et Biophysica Acta, 1319*, 283–292.

Bazer, F. W., Johnson, G. A., & Wu, G. (2015). Amino acids and conceptus development during the peri-implantation period of pregnancy. *Advances in Experimental Medicine and Biology, 843*, 23–52.

Belkacemi, L., Lash, G. E., Macdonald-Goodfellow, S. K., Caldwell, J. D., & Graham, C. H. (2005). Inhibition of human trophoblast invasiveness by high glucose concentrations. *The Journal of Clinical Endocrinology and Metabolism, 90*, 4846–4851.

Burton, G. J., & Fowden, A. L. (2012). Review: The placenta and developmental programming: Balancing fetal nutrient demands with maternal resource allocation. *Placenta, 33*(Suppl), S23–S27.

Cawyer, C. R., Horvat, D., Leonard, D., Allen, S. R., Jones, R. O., Zawieja, D. C., Kuehl, T. J., & Uddin, M. N. (2014). Hyperglycemia impairs cytotrophoblast function via stress signaling. *American Journal of Obstetrics and Gynecology, 211*, 541.e1–541.e8.

Chang, S.-C., & Vivian Yang, W.-C. (2013). Hyperglycemia induces altered expressions of angiogenesis associated molecules in the trophoblast. *Evidence-Based Complementary and Alternative Medicine, 2013*, 457971–457982.

Cleal, J. K., Day, P. E., Simner, C. L., Barton, S. J., Mahon, P. A., Inskip, H. M., Godfrey, K. M., Hanson, M. A., Cooper, C., Lewis, R. M., & Harvey, N. C. (2015). Placental amino acid transport may be regulated by maternal vitamin D and vitamin D-binding protein: Results from the Southampton Women's Survey. *British Journal of Nutrition, 113*, 1903–1910.

Correia-Branco, A., Azevedo, C. F., Araujo, J. R., Guimaraes, J. T., Faria, A., Keating, E., & Martel, F. (2015). Xanthohumol impairs glucose uptake by a human first-trimester extravillous trophoblast cell line (HTR-8/SVneo cells) and impacts the process of placentation. *Molecular Human Reproduction, 21*(10), 803–815.

Demir, R., Kayisli, U. A., Seval, Y., Celik-Ozenci, C., Korgun, E. T., Demir-Weusten, A. Y., & Huppertz, B. (2004). Sequential expression of VEGF and its receptors in human placental villi during very early pregnancy: Differences between placental vasculogenesis and angiogenesis. *Placenta, 25*, 560–572.

Desforges, M., Greenwood, S. L., Glazier, J. D., Westwood, M., & Sibley, C. P. (2010). The contribution of SNAT1 to system A amino acid transporter activity in human placental trophoblast. *Biochemical and Biophysical Research Communications, 398*, 130–134.

Di Simone, N., Di Nicuolo, F., Marzioni, D., Castellucci, M., Sanguinetti, M., D'Lppolito, S., & Caruso, A. (2009). Resistin modulates glucose uptake and glucose transporter-1 (GLUT-1) expression in trophoblast cells. *Journal of Cellular and Molecular Medicine, 13*, 388–397.

Dilworth, M. R., & Sibley, C. P. (2013). Review: Transport across the placenta of mice and women. *Placenta, 34*(Suppl), S34–S39.

Economides, D. L., Nicolaides, K. H., & Campbell, S. (1990). Relation between maternal-to-fetal blood glucose gradient and uterine and umbilical Doppler blood flow measurements. *British Journal of Obstetrics and Gynaecology, 97*, 543–544.

Ericsson, A., Hamark, B., Jansson, N., Johansson, B. R., Powell, T. L., & Jansson, T. (2005). Hormonal regulation of glucose and system A amino acid transport in first trimester placental villous fragments. *American Journal of Physiology. Regulatory, Integrative and Comparative Physiology, 288*, R656–R662.

Esterman, A., Greco, M. A., Mitani, Y., Finlay, T. H., Ismail-Beigi, F., & Dancis, J. (1997). The effect of hypoxia on human trophoblast in culture: Morphology, glucose transport and metabolism. *Placenta, 18*, 129–136.

Ferreira, R. V., Gombar, F. M., da Silva Faria, T., Costa, W. S., Sampaio, F. J., & da Fonte Ramos, C. (2010). Metabolic programming of ovarian angiogenesis and folliculogenesis by maternal malnutrition during lactation. *Fertility and Sterility, 93*, 2572–2580.

Frolova, A., Flessner, L., Chi, M., Kim, S. T., Foyouzi-Yousefi, N., & Moley, K. H. (2009). Facilitative glucose transporter type 1 is differentially regulated by progesterone and estrogen in murine and human endometrial stromal cells. *Endocrinology, 150*, 1512–1520.

Frolova, A. I., & Moley, K. H. (2011). Quantitative analysis of glucose transporter mRNAs in endometrial stromal cells reveals critical role of GLUT1 in uterine receptivity. *Endocrinology, 152*, 2123–2128.

Fukushima, K., Miyamoto, S., Tsukimori, K., Kobayashi, H., Seki, H., Takeda, S., Kensuke, E., Ohtani, K., Shibuya, M., & Nakano, H. (2005). Tumor necrosis factor and vascular endothelial growth factor induce endothelial integrin repertories, regulating endovascular differentiation and apoptosis in a human extravillous trophoblast cell line. *Biology of Reproduction, 73*, 172–179.

Gheorghe, C. P., Goyal, R., Holweger, J. D., & Longo, L. D. (2009). Placental gene expression responses to maternal protein restriction in the mouse. *Placenta, 30*, 411–417.
Gordon, M. C., Zimmerman, P. D., Landon, M. B., Gabbe, S. G., & Kniss, D. A. (1995). Insulin and glucose modulate glucose transporter messenger ribonucleic acid expression and glucose uptake in trophoblasts isolated from first-trimester chorionic villi. *American Journal of Obstetrics and Gynecology, 173*, 1089–1097.
Gude, N. M., Stevenson, J. L., Rogers, S., Best, J. D., Kalionis, B., Huisman, M. A., Erwich, J. J., Timmer, A., & King, R. G. (2003). GLUT12 expression in human placenta in first trimester and term. *Placenta, 24*, 566–570.
Hahn, T., Barth, S., Weiss, U., Mosgoeller, W., & Desoye, G. (1998). Sustained hyperglycemia in vitro down-regulates the GLUT1 glucose transport system of cultured human term placental trophoblast: A mechanism to protect fetal development? *FASEB Journal, 12*, 1221–1231.
Hahn, T., Hahn, D., Blaschitz, A., Korgun, E. T., Desoye, G., & Dohr, G. (2000). Hyperglycaemia-induced subcellular redistribution of GLUT1 glucose transporters in cultured human term placental trophoblast cells. *Diabetologia, 43*, 173–180.
Hayashi, M., Sakata, M., Takeda, T., Yamamoto, T., Okamoto, Y., Sawada, K., Kimura, A., Minekawa, R., Tahara, M., Tasaka, K., & Murata, Y. (2004). Induction of glucose transporter 1 expression through hypoxia-inducible factor 1alpha under hypoxic conditions in trophoblast-derived cells. *Journal of Endocrinology, 183*, 145–154.
Henriksen, T., & Clausen, T. (2002). The fetal origins hypothesis: Placental insufficiency and inheritance versus maternal malnutrition in well-nourished populations. *Acta Obstetricia et Gynecologica Scandinavica, 81*, 112–114.
Illsley, N. P. (2000). Glucose transporters in the human placenta. *Placenta, 21*, 14–22.
Illsley, N. P., Sellers, M. C., & Wright, R. L. (1998). Glycaemic regulation of glucose transporter expression and activity in the human placenta. *Placenta, 19*, 517–524.
Jansson, N., Rosario, F. J., Gaccioli, F., Lager, S., Jones, H. N., Roos, S., Jansson, T., & Powell, T. L. (2013). Activation of placental mTOR signaling and amino acid transporters in obese women giving birth to large babies. *Journal of Clinical Endocrinology and Metabolism, 98*, 105–113.
Jansson, T., Wennergren, M., & Illsley, N. P. (1993). Glucose transporter protein expression in human placenta throughout gestation and in intrauterine growth retardation. *Journal of Clinical Endocrinology and Metabolism, 77*, 1554–1562.
Jansson, T., Ylven, K., Wennergren, M., & Powell, T. L. (2002). Glucose transport and system A activity in syncytiotrophoblast microvillous and basal plasma membranes in intrauterine growth restriction. *Placenta, 23*, 392–399.
Janzen, C., Lei, M. Y., Cho, J., Sullivan, P., Shin, B. C., & Devaskar, S. U. (2013). Placental glucose transporter 3 (GLUT3) is up-regulated in human pregnancies complicated by late-onset intrauterine growth restriction. *Placenta, 34*, 1072–1078.
Jauniaux, E., Gulbis, B., Gerlo, E., & Rodeck, C. (1998). Free amino acid distribution inside the first trimester human gestational sac. *Early Human Development, 51*, 159–169.
Khalil, A., Hardman, L., & O Brien, P. (2015). The role of arginine, homoarginine and nitric oxide in pregnancy. *Amino Acids, 47*, 1715–1727.
Kleppa, M. J., Erlenwein, S. V., Darashchonak, N., von Kaisenberg, C. S., & von Versen-Hoynck, F. (2014). Hypoxia and the anticoagulants dalteparin and acetylsalicylic acid affect human placental amino acid transport. *PLoS One, 9*, e99217.
Korgun, E. T., Celik-Ozenci, C., Seval, Y., Desoye, G., & Demir, R. (2005). Do glucose transporters have other roles in addition to placental glucose transport during early pregnancy? *Histochemistry and Cell Biology, 123*, 621–629.
Lagarde, M., Liu, M., Vericel, E., Calzada, C., Chen, P., Driss, F., & Guichardant, M. (2014). Docosahexaenoic acid, protectin synthesis: Relevance against atherothrombogenesis. *Proceedings of the Nutrition Society, 73*, 186–189.

Lager, S., Jansson, T., & Powell, T. L. (2014). Differential regulation of placental amino acid transport by saturated and unsaturated fatty acids. *American Journal of Physiology – Cellular Physiology, 307*, C738–C744.

Luther, J., Milne, J., Aitken, R., Matsuzaki, M., Reynolds, L., Redmer, D., & Wallace, J. (2007). Placental growth, angiogenic gene expression, and vascular development in undernourished adolescent sheep. *Biology of Reproduction, 77*, 351–357.

Ogura, K., Sakata, M., Yamaguchi, M., Kurachi, H., & Murata, Y. (1999). High concentration of glucose decreases glucose transporter-1 expression in mouse placenta in vitro and in vivo. *Journal of Endocrinology, 160*, 443–452.

Ramakrishnan, U., Grant, F., Goldenberg, T., Zongrone, A., & Martorell, R. (2012). Effect of women's nutrition before and during early pregnancy on maternal and infant outcomes: A systematic review. *Paediatric and Perinatal Epidemiology, 26*(Suppl 1), 285–301.

Rosario, F. J., Jansson, N., Kanai, Y., Prasad, P. D., Powell, T. L., & Jansson, T. (2011). Maternal protein restriction in the rat inhibits placental insulin, mTOR, and STAT3 signaling and down-regulates placental amino acid transporters. *Endocrinology, 152*, 1119–1129.

Tsujii, H., Khandoker, M. A. M. Y., & Hamano, K.-I. (2001). Lipid in mammalian embryo development. *Journal of Mammalian Ova Research, 18*, 73–80.

von Wolff, M., Ursel, S., Hahn, U., Steldinger, R., & Strowitzki, T. (2003). Glucose transporter proteins (GLUT) in human endometrium: Expression, regulation, and function throughout the menstrual cycle and in early pregnancy. *Journal of Clinical Endocrinology and Metabolism, 88*, 3885–3892.

Weiss, U., Cervar, M., Puerstner, P., Schmut, O., Haas, J., Mauschitz, R., Arikan, G., & Desoye, G. (2001). Hyperglycaemia in vitro alters the proliferation and mitochondrial activity of the choriocarcinoma cell lines BeWo, JAR and JEG-3 as models for human first-trimester trophoblast. *Diabetologia, 44*, 209–219.

Zamudio, S., Torricos, T., Fik, E., Oyala, M., Echalar, L., Pullockaran, J., Tutino, E., Martin, B., Belliappa, S., Balanza, E., & Illsley, N. P. (2010). Hypoglycemia and the origin of hypoxia-induced reduction in human fetal growth. *PLoS One, 5*, e8551.

Chapter 4
Dietary Fatty Acids and Placentation

Contents

4.1 Introduction

Essential fatty acids (EFAs) belong to the $n-6$ (omega-6) and $n-3$ (omega-3) families, starting with the precursors; linoleic acid (LA), 18:2$n-6$; and alpha-linolenic acid (ALA), 18:3$n-3$. The $n-6$ series of fatty acids contain two or more double bonds, with the first double bond on the sixth carbon from the methyl end of the molecule; the $n-3$ fatty acids contain three or more double bonds, with the first double bond on the third carbon atom from the methyl end. $N-3$ fatty acids and $n-6$ fatty acids play crucial biological roles that include structure and function of cell membranes, providing substrates for the production of signaling molecules such as eicosanoids and modulating expression of genes involved in cell homeostasis (Smith 1989; Innis 1991; Dutta-Roy 1994). Dietary LA can be converted to AA in the body, mostly in the liver. The three main $n-3$ fatty acids are ALA, docosahexaenoic acid (DHA), 22:6$n-3$, and eicosapentaenoic acid (EPA), 20:5$n-3$. Through the same desaturase and elongase enzymes, the $n-3$ fatty acids, ALA can be converted into EPA and DHA. The enzymes responsible for the metabolism of both $n-6$ fatty acids and $n-3$ fatty acids are the COX, lipoxygenases (LOXs), and cytochrome P450 (CYP 450). Several eicosanoids derived from the $n-6$ fatty acids promote tumor angiogenesis, such as the prostaglandins (PGH_2, PGE_2, PGI_2), leukotrienes (4-series LTs), thromboxanes (TXA_2), and hydroxyeicosatraenoic acids (12-HETE,

A.K. Duttaroy, S. Basak, *Early Nutrition and Lifestyle Factors*,
DOI 10.1007/978-3-319-38804-5_4

15-HETE) (Smith 1989; Nie et al. 2000; Hoagland et al. 2001; Pai et al. 2001; Bagga et al. 2003; Pola et al. 2004; Jin et al. 2009; Kamiyama et al. 2006). These eicosanoids make the tumor microenvironment more favorable for neoplasms and metastasis by encouraging the transcription of angiogenic growth factors, increasing the rate of endothelial cell migration and proliferation, and increasing the rate of vascularization. In contrast, $n-3$ fatty acid metabolism produces leukotrienes and prostaglandins that attenuate excess vascularization. $N-3$ and $n-6$ fatty acids compete each other for incorporation into the cell membrane in addition to enzymes for eicosanoid production, including COX-2 and 5-LOX. Thus, high levels of tissue $n-3$ fatty acids can reduce angiogenesis through decreased production of pro-angiogenic AA-derived eicosanoids, membrane receptor-ligand interactions, and through $n-3$ fatty acid's intrinsic antitumor properties (Nie et al. 2000; Pai et al. 2001; Bagga et al. 2003; Pola et al. 2004; Kamiyama et al. 2006; Yuan et al. 2009). Furthermore, $n-3$ fatty acids have been found to downregulate expression of angiogenic growth factors such as VEGF, PDGF, IL-6, and MMP-2 (Kang and Weylandt 2008; Spencer et al. 2009). $N-3$ fatty acids are only found in marine fish and certain vegetables and nuts, whereas corn and soybean oils, processed foods containing these oils, and grain-fed meat contain high levels of $n-6$ fatty acids. Now, the ratio of $n-6/n-3$ fatty acids is approximately 15:1 or higher; our bodies may not be accustomed to utilizing such high levels of $n-6$ fatty acids. This is considered to be one of many factors responsible for the relatively recent rise in chronic diseases, predominantly those associated with inflammation including cancer, heart disease, arthritis, and diabetes (Simopoulos 2002). The maternal, fetal, and neonatal EFA/LCFA status is an important determinant of health and disease in infancy and later life (Innis 1991; Uauy et al. 2001). In fact, fetal brain and the retina are very rich in AA and DHA (Innis 2007). The numerous studies have demonstrated a positive effect of supplementation with DHA in pregnant women in terms of less premature birth and in the child; in terms of complex brain performance, like visual acuity, attention spans, and intelligence; and in terms of mother's overall health. In addition, DHA may reduce the incidence of preeclampsia and postpartum depression by stimulating placental angiogenesis.

In general, $n-3$ fatty acids have anti-inflammatory and anticancer effects, whereas $n-6$ fatty acids promote inflammation and carcinogenesis (Dutta-Roy 2000; Sterescu et al. 2006; Sapieha et al. 2011). $N-3$ long-chain polyunsaturated fatty acids (LCFAs) such as eicosapentaenoic acid (EPA), 20:5$n-3$, and docosahexaenoic acid (DHA), 22:6$n-3$, inhibit angiogenesis, whereas $n-6$ LCFA such as arachidonic acid (AA), 20:4$n-6$, promotes angiogenesis (Belury 2002; Sterescu et al. 2006; Spencer et al. 2009; Chen 2010; Sapieha et al. 2011). These fatty acids influence angiogenesis via several mechanisms such as expression of angiogenic factors, VEGF, ANGPTL4, and other modulators such as eicosanoids, cyclooxygenase (COX), fatty acid-binding proteins (FABPs), and nitric oxide (NO) (Spencer et al. 2009).

Since supplementation of $n-3$ fatty acids is recommended in pregnancy for optimal feto-placental growth and development (Innis 1991; Uauy et al. 2001), it is therefore important to investigate the effects of these fatty acids in placental angiogenesis. With the recent spate of clinical work on regulators of angiogenesis, these observations lead us to believe that the regulation of placental angiogenesis could become a novel and powerful method for ensuring positive outcomes for most pregnancies. This chapter deals with effects of fatty acids on placental angiogenesis.

4.2 Modulation of Angiogenesis by LCFAs

Both $n-3$ and $n-6$ fatty acids are involved directly or indirectly in angiogenesis (Hardman 2004; Spencer et al. 2009; Salvado et al. 2012). Eicosanoids produced from AA stimulate angiogenesis, whereas those produced from EPA and DHA inhibit angiogenesis and tumorigenesis (Hardman 2004; Spencer et al. 2009; Salvado et al. 2012). In marked contrast to the effects observed with the $n-3$ LCFAs, the $n-6$ LCFAs have stimulatory or neutral effect on angiogenic processes such as tube formation, cell migration, or cell proliferation (Hardman 2004; Szymczak et al. 2008). The expression of enzymes involved in the biosynthesis of eicosanoids notably COX-2, 5-LOX (lipoxygenase), and 12-LOX, is upregulated during tumor initiation and progression. COX-2 contributes to neovascularization, which is essential for tumor development. Overexpression of COX-2 in colon carcinoma cells increases their angiogenicity, as shown by an increased ability to stimulate endothelial cell migration and tube formation by producing prostaglandins and inducing pro-angiogenic factors such as VEGF and basic fibroblast growth factor (bFGF). In addition, inflammatory cells infiltrating the tumor tissue (e.g., macrophages and neutrophils) may release pro-inflammatory cytokines such as IL-1β or TNFα that can induce angiogenesis, an effect partially attributable to increased COX-2 expression in tumor, stromal, and vascular endothelial cells. The anti-angiogenic activity of $n-3$ LCFAs is mediated usually via inhibition of production of AA-derived eicosanoids. LCFAs and their derivatives are reported to modulate several angiogenic factors such as VEGF, ANGPTL4, PDGF, leptin, and TNFα (Spencer et al. 2009; Salvado et al. 2012). Numerous studies have demonstrated that VEGF or its receptors are upregulated in many tumors (Takahashi 2011). $N-3$ fatty acids affect expression of pro-angiogenic factors (Massaro et al. 2010; Scoditti et al. 2010). EPA and DHA significantly suppress endothelial cell proliferation, migration, and tube formation (Kanayasu et al. 1991; Yang et al. 1998; Murota et al. 2000; Kim et al. 2005; Spencer et al. 2009). Downregulation of VEGF receptors by EPA is mediated via suppression of NFkB activation (Ghosh-Choudhury et al. 2009). In addition, EPA downregulates the expression of Flk-1 receptors in a dose-dependent manner and upregulates the expression of Flt1 receptors (Yang et al. 1998). The tube formation induced by VEGF was suppressed by $n-3$ LCFAs via downregulation of VEGFR-2 in the endothelial cells (Tsuji et al. 2003; Calviello et al. 2004; Szymczak et al. 2008; Spencer et al. 2009). EPA and DHA have shown to have potent anti-angiogenic effects in cancer cells by inhibiting the production of many important angiogenic mediators such as VEGF, PDGF, COX-2, PGE_2, and nitric oxide (NO). 4-hydroxy-DHA, a 5-lipoxygenase product of DHA, was reported to inhibit endothelial cell proliferation and sprouting angiogenesis via PPARγ (Sapieha et al. 2011). In addition, both EPA and DHA decrease the levels of O_2^- and thereby reduce reactive oxygen species (ROS)-mediated angiogenesis (Matesanz et al. 2010; Maraldi et al. 2010). $N-3$ LCFAs can increase NO production by displacing eNOS from the caveolar fraction (Li et al. 2007). Increased NO levels by these fatty acids may decrease VEGFR signal-mediated angiogenesis (Matesanz et al. 2010). Indeed, a high ratio of O_2^-/NO is implicated as a contributory factor in disturbed angiogenesis in diabetes (Matesanz et al. 2010; Hamed et al. 2011). This report summarized that $n-6$ fatty acids increase, whereas $n-3$ fatty acids decrease angiogenesis. Since tumor progression and metastasis depend on angiogenesis, reducing the tissue $n-6/n-3$ fatty acids may be beneficial in cancers.

4.3 FABPs and Their Roles in Angiogenesis

FABPs, ubiquitously expressed in tissues including the placenta, may play a role in intracellular fatty acid transport and metabolism (Campbell et al. 1998; Duttaroy 2009). The transport properties of the FABPs are governed in part by specific protein-protein and protein-membrane interactions, and the helix-turn-helix domain of the FABPs is critical although likely not exclusive in specifying these interactions (Storch and Thumser 2010). Several of the FABPs have been shown to deliver their ligands to nuclear transcription factors, thus modulating gene expression in a tissue-specific manner. FABPs may be envisioned as central regulators of lipid disposition at the cell and tissue levels that have a profound impact on any processes involving fatty acid-induced angiogenesis. Recent data suggested that certain FABPs have a role in angiogenesis. In fact, FABP expression is widely regulated by various pro-angiogenic mediators (Mousiolis et al. 2012). Among FABPs, FABP-4 originally identified as an adipocyte-specific protein promotes proliferation of endothelial cells. FABP-4 mRNA and protein levels were significantly induced in cultured endothelial cells by VEGF and bFGF treatment. The effect of VEGF-A on FABP-4 expression was inhibited by siRNA-mediated knockdown of VEGFR-2, whereas the VEGFR-1 agonists, PIGF1 and PIGF2, had no such effect. Thus, FABP-4 emerged as a novel target of the VEGF/VEGFR-2 pathway and a positive regulator of cell proliferation and angiogenesis in endothelial cells (Elmasri et al. 2009, 2012b; Cataltepe et al. 2011; Ghelfi et al. 2013). FABP-4 was expressed in an angiogenesis-dependent pathology, infantile hemangioma, being the most common tumor of infancy and endothelial cells (Fishman and Mulliken 1993). FABP-4 has a role on activation of several mitogenic pathways and expression of several key mediators of angiogenesis (Elmasri et al. 2012a). The expression of FABPs was regulated by fatty acids and leptin in the first-trimester trophoblast cells, HTR8/SVneo. c9,t11-CLA, an isomer of conjugated linoleic acid, specifically increased FABP-4 expression by tenfold compared to t10,c12-CLA. The c9,t11-CLA-induced FABP-4 expression was associated with tube formation and increased amount of ANGPTL4 secretion (Basak and Duttaroy 2013a). Leptin, a potent angiogenic growth factor, stimulates mRNA expression of FABP-4 along with several pro-angiogenic factors in first-trimester placental cells (Basak and Duttaroy 2012). FABP-3 and its expression correlate with inhibition of α1-integrin activity, cell adhesion, and invasion in breast cancer cells (Kusakari et al. 2006; Nevo et al. 2010). Further, expression of FABP-5 has been shown to modulate MMP-9 production and regulates invasive property of oral cancer cells (Fang et al. 2010). Despite similar ligand binding characteristics and highly homologous tertiary structures, each FABP appears to have unique functions in specific tissues (Storch and Thumser 2010). Recent data demonstrated that maternal serum FABP-4 is independently associated (Yan et al. 2016) with the subsequent development of preeclampsia. Elevated maternal serum FABP-4 levels may also play a role in the pathogenesis of preeclampsia through pathways related to insulin resistance, inflammation, and abnormal lipid metabolism. Increasing evidence suggests the angiogenic role of fatty acid transport/binding proteins in different cell systems including first-trimester placental trophoblast (Ghelfi et al. 2013; Basak and Duttaroy 2013a, b). Recently we determined the tube formation both at the basal and stimulated levels in the absence and presence of inhibitors of FABP-4 and VEGF signaling pathways. Basal level of tube formation was maximally decreased in the presence of FABP-4 inhibitor compared with the other VEGF signaling pathway inhibitors such as P38/MAPK and L-NAME, whereas DHA- and VEGF-induced tube formation was maximally inhibited by p38 MAP kinase inhibitor (63.7 % and 34.5 %, respectively) (Pandya et al. 2016). ANGPTL4- and oleic acid (OA)-induced tube formation was not inhibited by any of these inhibitors. The FABP-4 inhibitor also attenuated cell growth but did not affect ^{14}C fatty acid uptake. FABP-4 therefore may be involved in part in the basal level and stimulated tube formation by VEGF, DHA, and leptin, whereas it has little or no effect in ANGPTL4- and OA-induced tube formation in these cells. Thus, FABP-4 may play a differential role in fatty acid- and angiogenic growth factor-mediated tube formation in the first-trimester trophoblast cells in vitro. FABP-4 may not be involved as the key regulator in first-trimester placental trophoblasts. All these data suggest angiogenesis and lipid metabolism processes are interlinked. Modulators such as fatty acids can regulate both these processes together and control the balance between pro-/anti-angiogenic process that ultimately drives the feto-placental growth and development during early placentation (Fig. 4.1).

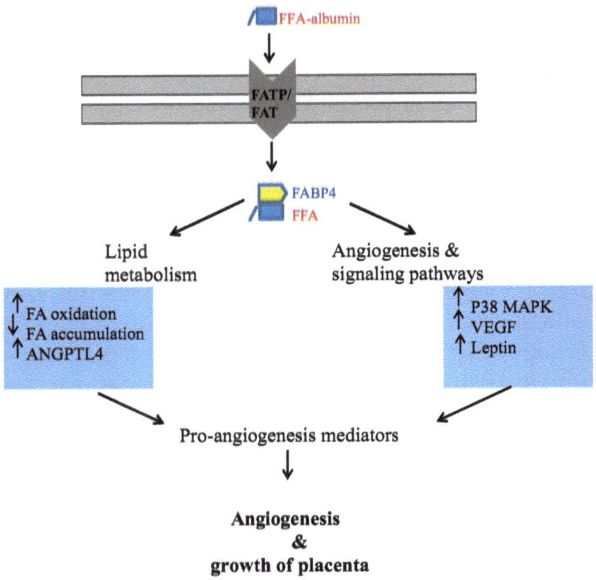

Fig. 4.1 Possible effects of fatty acids on metabolic and angiogenic modulators of the first-trimester placenta. FA, long-chain fatty acids

4.4 Fatty Acid Mediated-Angiogenesis in Human First-Trimester Placental Trophoblast Cells: Possible Mechanisms

During the third trimester of pregnancy, DHA requirements increase to support fetal growth, particularly of the brain and eyes (Innis 1991). DHA was taken up preferentially compared with other fatty acids by the last-trimester placental trophoblasts. DHA was also taken up by the first-trimester placental trophoblasts (Basak and Duttaroy 2013b). DHA increased tube formation to the greatest extent as compared to other long-chain fatty acids by increasing capillary tube length and tube numbers of extravillous trophoblast cells, HTR8/SVneo. DHA stimulated tube formation via expression of the most potent angiogenic factor, VEGF, in extravillous trophoblast cells (Johnsen et al. 2011). DHA may also help early placentation process by increasing angiogenesis (Johnsen et al. 2011). This is in contrast with the generally observed inhibitory effects of $n-3$ LCFAs on angiogenesis in many cell types including tumors (Spencer et al. 2009). The mechanism responsible for the increased expression of VEGF in placental trophoblast cells by DHA is not known at present. Expression of VEGF by DHA however is very unique as its mRNA is induced by a variety of growth factors and cytokines, including PDGF, EGF, TNFα, TGF-β1, and IL-1β, but not by any fatty acids. DHA-induced VEGF expression was not accompanied with the expression of COX-2 and HIF1α

Table 4.1 Effect of long-chain fatty acids on the mRNA expression of angiogenic growth factors in HTR8/SVneo cells

Long-chain fatty acids	Angiogenic factors mRNA level (gene/TBP)	
	VEGF	ANGPTL4
Control	1.0	1.0
OA	0.96	8.5
t-9 ELA	0.76	1.95
AA	1.2	18*
c9,t11-CLA	1.13	12.43*
EPA	1.9	20.0*
DHA	2.1*	23.7*

*$p < 0.05$ vs. control [Adapted from Basak and Duttaroy (2013b)]

genes in these cells (Johnsen et al. 2011), indicating that DHA metabolites per se may not be involved in the VEGF expression. Since the PPARγ ligand did not stimulate VEGF expression in these cells, it is unlikely that DHA stimulation of VEGF expression involves PPARγ (Gottlicher et al. 1993). DHA may also alter phosphorylation events in native mRNA processing, mRNA transport and stabilization, and mRNA degradation rates (Salem et al. 2001; Uauy et al. 2001). DHA stimulates VEGF expression and secretion, whereas fatty acids such as EPA, ARA, OA, and CLA promote ANGPTL4 secretion without affecting VEGF synthesis in the placental trophoblast cells (Table 4.1). These data indicate the different mechanisms of action of DHA compared with other fatty acids in angiogenic process (Johnsen et al. 2011). The mechanisms that determine the angiogenic capacity of DHA compared with other fatty acids may underlie important differences in the mechanism of actions of these structurally different fatty acids. Based on available data, we postulate that free fatty acids such as c9t11-CLA, EPA, DHA, and leptin stimulate expression of FABP-4 that has been demonstrated as a target protein for VEGF (Fig. 4.2). First-trimester trophoblast cells, HTR8/SVneo, express and secrete VEGF significantly in the presence of fatty acids, and FABP-4 could be a mediator of this association of VEGF, ANGPTL4, and FABP-4. The mechanism responsible for increased secretion of ANGPTL4 in placental trophoblast cells by fatty acids other than DHA is yet to be resolved. More detailed work is however required for understanding the mechanisms involved in fatty acid-stimulated angiogenesis. Future work should explore the roles of ANGPTL4, VEGF, and FABP-4 in first-trimester human placental trophoblasts and their involvement in angiogenesis. cis-9,trans-11(c9t11)-conjugated linoleic acid (CLA) increased tube formation (as a measure of angiogenesis) in first-trimester human placental cells mediated via ANGPTL4 (Johnsen et al. 2011; Basak and Duttaroy 2013a). In first-trimester placental trophoblast cells, the effects of fatty acids were quite opposite to what was observed in case tumors; LCFAs induce VEGF and ANGPTL4 expression (Johnsen et al. 2011; Kang and Liu 2013; Basak and Duttaroy 2013a, b).

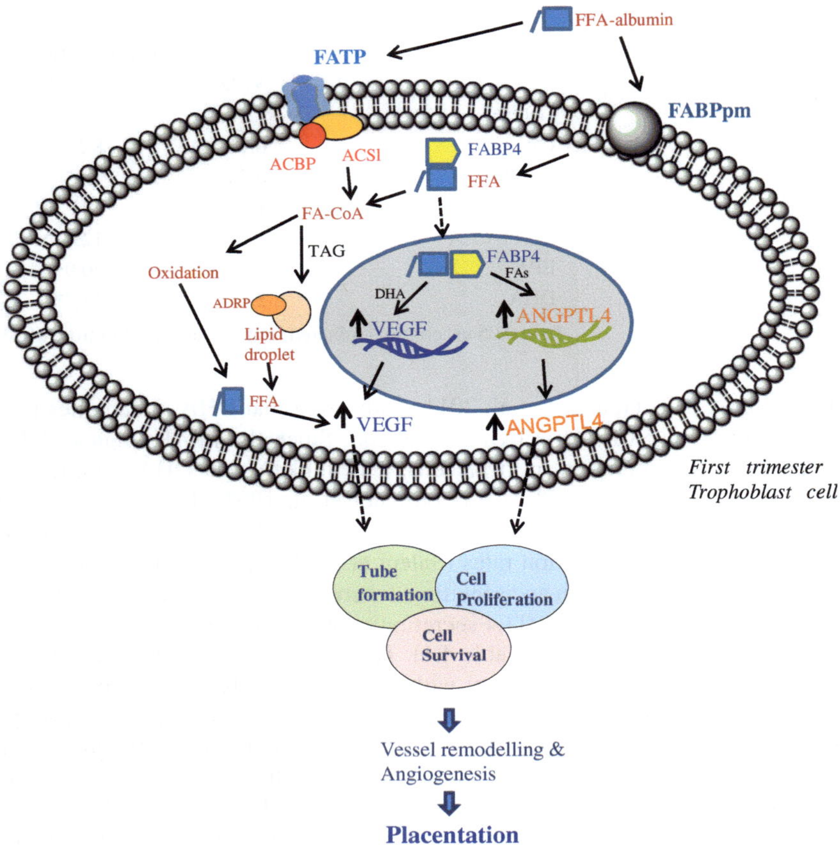

Fig. 4.2 Effects of long-chain fatty acids on the growth and angiogenesis factors of the first-trimester placenta. Possible contribution of these fatty acids in modulating angiogenic growth factors such as VEGF and ANGPTL4 is depicted. *FFA* free fatty acids, *FATP* fatty acid transporter protein, *pFABPpm* plasma membrane fatty acid-binding protein, *ACSL* acyl-coA synthetase, *ADRP* adipose differentiation-related protein, *ACBP* acyl-coA-binding protein, *TAG* triacylglycerol, *DHA* docosahexaenoic acid, *FAs* fatty acids, *FABP-4* fatty acid-binding protein 4

4.5 Summary

Early placental angiogenesis is critical for the establishment of the placental vascularization and thus for normal fetal growth and development. Inadequate placental vascular development may compromise the feto-placental growth and development. It seems reasonable to suggest that placental vascular defects and placental dysfunction may be an important cause of infertility to the mother and fetal growth retardation. With the recent spate of studies on regulators of angiogenesis, these observations lead us to believe that regulation of placental angiogenesis could become a novel and powerful way for ensuring positive outcomes for most pregnancies. DHA and other $n-6$ fatty acids including CLA stimulate angiogenesis in first-trimester placental cells. These dietary fatty acids not only stimulated the expression of major angiogenic factors such as VEGF and ANGPTL4 but also stimulated expression of FABP-4 and FABP-3 which are known to directly modulate angiogenesis. FABP-4 may be involved in part in the basal level and stimulated tube formation by VEGF, DHA, and leptin, whereas it has little or no effect in ANGPTL4- and OA-induced tube formation in these cells. Thus, FABP-4 may play a differential role in fatty acid- and angiogenic growth factor-mediated tube formation in the first-trimester trophoblast cells in vitro.

References

Bagga, D., Wang, L., Farias-Eisner, R., Glaspy, J. A., & Reddy, S. T. (2003). Differential effects of prostaglandin derived from omega-6 and omega-3 polyunsaturated fatty acids on COX-2 expression and IL-6 secretion. *Proceedings of the National Academy of Sciences of the United States of America, 100*, 1751–1756.

Basak, S., & Duttaroy, A. K. (2012). Leptin induces tube formation in first-trimester extravillous trophoblast cells. *European Journal of Obstetrics, Gynecology, and Reproductive Biology, 164* (1), 24–29.

Basak, S., & Duttaroy, A. K. (2013a). cis-9, trans-11 conjugated linoleic acid stimulates expression of angiopoietin like-4 in the placental extravillous trophoblast cells. *Biochimica et Biophysica Acta, 1831*, 834–843.

Basak, S., & Duttaroy, A. K. (2013b). Effects of fatty acids on angiogenic activity in the placental extravillous trophoblast cells. *Prostaglandins, Leukotrienes, and Essential Fatty Acids, 88*, 155–162.

Belury, M. A. (2002). Dietary conjugated linoleic acid in health: Physiological effects and mechanisms of action. *Annual Review of Nutrition, 22*, 505–531.

Calviello, G., DI Nicuolo, F., Gragnoli, S., Piccioni, E., Serini, S., Maggiano, N., Tringali, G., Navarra, P., Ranelletti, F. O., & Palozza, P. (2004). n-3 PUFAs reduce VEGF expression in human colon cancer cells modulating the COX-2/PGE2 induced ERK-1 and -2 and HIF-1alpha induction pathway. *Carcinogenesis, 25*, 2303–2310.

Campbell, F. M., Bush, P. G., Veerkamp, J. H., & Dutta-Roy, A. K. (1998). Detection and cellular localization of plasma membrane-associated and cytoplasmic fatty acid-binding proteins in human placenta. *Placenta, 19*, 409–415.

Cataltepe, O., Arikan, M. C., Ghelfi, E., Karaaslan, C., Ozsurekci, Y., Dresser, K., Li, Y., Smith, T. W., & Cataltepe, S. (2011). Fatty acid binding protein 4 is expressed in distinct endothelial and non-endothelial cell populations in glioblastoma. *Neuropathology and Applied Neurobiology, 38*(5), 400–10.

Chen, J. (2010). Src may be involved in the anti-cancer effect of conjugated linoleic acid. Comment on: CLA reduces breast cancer cell growth and invasion through ER and PI3K/ Akt pathways. *Chemico-Biological Interactions, 186*, 250–251. author reply 252–253.

Duttaroy, A. K. (2009). Transport of fatty acids across the human placenta: A review. *Progress in Lipid Research, 48*, 52–61.

Dutta-Roy, A. K. (1994). Insulin mediated processes in platelets, erythrocytes and monocytes/ macrophages: Effects of essential fatty acid metabolism. *Prostaglandins, Leukotrienes, and Essential Fatty Acids, 51*, 385–399.

Dutta-Roy, A. K. (2000). Transport mechanisms for long-chain polyunsaturated fatty acids in the human placenta. *American Journal of Clinical Nutrition, 71*, 315S–322S.

Elmasri, H., Ghelfi, E., Yu, C. W., Traphagen, S., Cernadas, M., Cao, H., Shi, G. P., Plutzky, J., Sahin, M., Hotamisligil, G., & Cataltepe, S. (2012a). Endothelial cell-fatty acid binding protein 4 promotes angiogenesis: Role of stem cell factor/c-kit pathway. *Angiogenesis, 15*(3), 457–68.

Elmasri, H., Ghelfi, E., Yu, C. W., Traphagen, S., Cernadas, M., Cao, H., Shi, G. P., Plutzky, J., Sahin, M., Hotamisligil, G., & Cataltepe, S. (2012b). Endothelial cell-fatty acid binding protein 4 promotes angiogenesis: Role of stem cell factor/c-kit pathway. *Angiogenesis, 15*, 457–468.

Elmasri, H., Karaaslan, C., Teper, Y., Ghelfi, E., Weng, M., Ince, T. A., Kozakewich, H., Bischoff, J., & Cataltepe, S. (2009). Fatty acid binding protein 4 is a target of VEGF and a regulator of cell proliferation in endothelial cells. *The FASEB Journal, 23*, 3865–3873.

Fang, L. Y., Wong, T. Y., Chiang, W. F., & Chen, Y. L. (2010). Fatty-acid-binding protein 5 promotes cell proliferation and invasion in oral squamous cell carcinoma. *Journal of Oral Pathology and Medicine, 39*, 342–348.

Fishman, S. J., & Mulliken, J. B. (1993). Hemangiomas and vascular malformations of infancy and childhood. *Pediatric Clinics of North America, 40*, 1177–1200.

Ghelfi, E., Yu, C. W., Elmasri, H., Terwelp, M., Lee, C. G., Bhandari, V., Comhair, S. A., Erzurum, S. C., Hotamisligil, G. S., Elias, J. A., & Cataltepe, S. (2013). Fatty acid binding protein 4 regulates VEGF-induced airway angiogenesis and inflammation in a transgenic mouse model: Implications for asthma. *The American Journal of Pathology, 182*, 1425–1433.

Ghosh-Choudhury, T., Mandal, C. C., Woodruff, K., St Clair, P., Fernandes, G., Choudhury, G. G., & Ghosh-Choudhury, N. (2009). Fish oil targets PTEN to regulate NFkappaB for downregulation of anti-apoptotic genes in breast tumor growth. *Breast Cancer Research and Treatment, 118*, 213–228.

Gottlicher, M., Demoz, A., Svensson, D., Tollet, P., Berge, R. K., & Gustafsson, J. A. (1993). Structural and metabolic requirements for activators of the peroxisome proliferator-activated receptor. *Biochemical Pharmacology, 46*, 2177–2184.

Hamed, E. A., Zakary, M. M., Abdelal, R. M., & Abdel Moneim, E. M. (2011). Vasculopathy in type 2 diabetes mellitus: Role of specific angiogenic modulators. *Journal of Physiology and Biochemistry, 67*, 339–349.

Hardman, W. E. (2004). (n-3) fatty acids and cancer therapy. *The Journal of Nutrition, 134*, 3427S–3430S.

Hoagland, K. M., Maier, K. G., Moreno, C., Yu, M., & Roman, R. J. (2001). Cytochrome P450 metabolites of arachidonic acid: Novel regulators of renal function. *Nephrology, Dialysis, Transplantation, 16*, 2283–2285.

Innis, S. M. (1991). Essential fatty acids in growth and development. *Progress in Lipid Research, 30*, 39–103.

Innis, S. M. (2007). Fatty acids and early human development. *Early Human Development, 83*, 761–766.

Jin, Y., Arita, M., Zhang, Q., Saban, D. R., Chauhan, S. K., Chiang, N., Serhan, C. N., & Dana, R. (2009). Anti-angiogenesis effect of the novel anti-inflammatory and pro-resolving lipid mediators. *Investigative Ophthalmology & Visual Science, 50*, 4743–4752.

Johnsen, G. M., Basak, S., Weedon-Fekjaer, M. S., Staff, A. C., & Duttaroy, A. K. (2011). Docosahexaenoic acid stimulates tube formation in first trimester trophoblast cells, HTR8/SVneo. *Placenta, 32*, 626–632.

Kamiyama, M., Pozzi, A., Yang, L., Debusk, L. M., Breyer, R. M., & Lin, P. C. (2006). EP2, a receptor for PGE2, regulates tumor angiogenesis through direct effects on endothelial cell motility and survival. *Oncogene, 25*, 7019–7028.

Kanayasu, T., Morita, I., Nakao-Hayashi, J., Asuwa, N., Fujisawa, C., Ishii, T., Ito, H., & Murota, S. (1991). Eicosapentaenoic acid inhibits tube formation of vascular endothelial cells in vitro. *Lipids, 26*, 271–276.

Kang, J. X., & Liu, A. (2013). The role of the tissue omega-6/omega-3 fatty acid ratio in regulating tumor angiogenesis. *Cancer and Metastasis Reviews, 32*, 201–210.

Kang, J. X., & Weylandt, K. H. (2008). Modulation of inflammatory cytokines by omega-3 fatty acids. *Subcellular Biochemistry, 49*, 133–143.

Kim, H. J., Vosseler, C. A., Weber, P. C., & Erl, W. (2005). Docosahexaenoic acid induces apoptosis in proliferating human endothelial cells. *Journal of Cellular Physiology, 204*, 881–888.

Kusakari, Y., Ogawa, E., Owada, Y., Kitanaka, N., Watanabe, H., Kimura, M., Tagami, H., Kondo, H., Aiba, S., & Okuyama, R. (2006). Decreased keratinocyte motility in skin wound on mice lacking the epidermal fatty acid binding protein gene. *Molecular and Cellular Biochemistry, 284*, 183–188.

Li, Q., Zhang, Q., Wang, M., Zhao, S., Ma, J., Luo, N., Li, N., Li, Y., Xu, G., & Li, J. (2007). Eicosapentaenoic acid modifies lipid composition in caveolae and induces translocation of endothelial nitric oxide synthase. *Biochimie, 89*, 169–177.

Maraldi, T., Prata, C., Caliceti, C., Vieceli Dalla Sega, F., Zambonin, L., Fiorentini, D., & Hakim, G. (2010). VEGF-induced ROS generation from NAD(P)H oxidases protects human leukemic cells from apoptosis. *International Journal of Oncology, 36*, 1581–1589.

Massaro, M., Scoditti, E., Carluccio, M. A., Campana, M. C., & DE Caterina, R. (2010). Omega-3 fatty acids, inflammation and angiogenesis: Basic mechanisms behind the cardioprotective effects of fish and fish oils. *Cellular and Molecular Biology, 56*, 59–82.

Matesanz, N., Park, G., McAllister, H., Leahey, W., Devine, A., Mcveigh, G. E., Gardiner, T. A., & McDonald, D. M. (2010). Docosahexaenoic acid improves the nitroso-redox balance and reduces VEGF-mediated angiogenic signaling in microvascular endothelial cells. *Investigative Ophthalmology and Visual Science, 51*, 6815–6825.

Mousiolis, A. V., Kollia, P., Skentou, C., & Messinis, I. E. (2012). Effects of leptin on the expression of fatty acid-binding proteins in human placental cell cultures. *Molecular Medicine Reports, 5*, 497–502.

Murota, S. I., Onodera, M., & Morita, I. (2000). Regulation of angiogenesis by controlling VEGF receptor. *Annals of the New York Academy of Sciences, 902*, 208–212. discussion 212–213.

Nevo, J., Mai, A., Tuomi, S., Pellinen, T., Pentikainen, O. T., Heikkila, P., Lundin, J., Joensuu, H., Bono, P., & Ivaska, J. (2010). Mammary-derived growth inhibitor (MDGI) interacts with integrin alpha-subunits and suppresses integrin activity and invasion. *Oncogene, 29*, 6452–6463.

Nie, D., Lamberti, M., Zacharek, A., Li, L., Szekeres, K., Tang, K., Chen, Y., & Honn, K. V. (2000). Thromboxane A(2) regulation of endothelial cell migration, angiogenesis, and tumor metastasis. *Biochemical and Biophysical Research Communications, 267*, 245–251.

Pai, R., Szabo, I. L., Soreghan, B. A., Atay, S., Kawanaka, H., & Tarnawski, A. S. (2001). PGE (2) stimulates VEGF expression in endothelial cells via ERK2/JNK1 signaling pathways. *Biochemical and Biophysical Research Communications, 286*, 923–928.

Pandya, A. D., Das, M. K., Sarkar, A., Vilasagaram, S., Basak, S., & Duttaroy, A. K. (2016). Tube formation in the first trimester placental trophoblast cells: Differential effects of angiogenic growth factors and fatty acids. *Cell Biology International, 40*(6), 652–661.

Pola, R., Gaetani, E., Flex, A., Aprahamian, T. R., Bosch-Marce, M., Losordo, D. W., Smith, R. C., & Pola, P. (2004). Comparative analysis of the in vivo angiogenic properties of stable prostacyclin analogs: A possible role for peroxisome proliferator-activated receptors. *Journal of Molecular and Cellular Cardiology, 36*, 363–370.

Salem, N., Jr., Litman, B., Kim, H. Y., & Gawrisch, K. (2001). Mechanisms of action of docosahexaenoic acid in the nervous system. *Lipids, 36*, 945–959.

Salvado, M. D., Alfranca, A., Haeggstrom, J. Z., & Redondo, J. M. (2012). Prostanoids in tumor angiogenesis: Therapeutic intervention beyond COX-2. *Trends in Molecular Medicine, 18*, 233–243.

Sapieha, P., Stahl, A., Chen, J., Seaward, M. R., Willett, K. L., Krah, N. M., Dennison, R. J., Connor, K. M., Aderman, C. M., Liclican, E., Carughi, A., Perelman, D., Kanaoka, Y., Sangiovanni, J. P., Gronert, K., & Smith, L. E. (2011). 5-Lipoxygenase metabolite 4-HDHA is a mediator of the antiangiogenic effect of omega-3 polyunsaturated fatty acids. *Science Translational Medicine, 3*, 69ra12.

Scoditti, E., Massaro, M., Carluccio, M. A., Distante, A., Storelli, C., & DE Caterina, R. (2010). PPARgamma agonists inhibit angiogenesis by suppressing PKCalpha- and CREB-mediated COX-2 expression in the human endothelium. *Cardiovascular Research, 86*, 302–310.

Simopoulos, A. P. (2002). Omega-3 fatty acids in inflammation and autoimmune diseases. *Journal of the American College of Nutrition, 21*, 495–505.

Smith, W. L. (1989). The eicosanoids and their biochemical mechanisms of action. *Biochemical Journal, 259*, 315–324.

Spencer, L., Mann, C., Metcalfe, M., Webb, M., Pollard, C., Spencer, D., Berry, D., Steward, W., & Dennison, A. (2009). The effect of omega-3 FAs on tumour angiogenesis and their therapeutic potential. *European Journal of Cancer, 45*, 2077–2086.

Sterescu, A. E., Rousseau-Harsany, E., Farrell, C., Powell, J., David, M., & Dubois, J. (2006). The potential efficacy of omega-3 fatty acids as anti-angiogenic agents in benign vascular tumors of infancy. *Medical Hypotheses, 66*, 1121–1124.

Storch, J., & Thumser, A. E. (2010). Tissue-specific functions in the fatty acid-binding protein family. *Journal of Biological Chemistry, 285*, 32679–32683.

Szymczak, M., Murray, M., & Petrovic, N. (2008). Modulation of angiogenesis by omega-3 polyunsaturated fatty acids is mediated by cyclooxygenases. *Blood, 111*, 3514–3521.

Takahashi, S. (2011). Vascular endothelial growth factor (VEGF), VEGF receptors and their inhibitors for antiangiogenic tumor therapy. *Biological & Pharmaceutical Bulletin, 34*, 1785–1788.

Tsuji, M., Murota, S. I., & Morita, I. (2003). Docosapentaenoic acid (22:5, n-3) suppressed tube-forming activity in endothelial cells induced by vascular endothelial growth factor. *Prostaglandins, Leukotrienes, and Essential Fatty Acids, 68*, 337–342.

Uauy, R., Hoffman, D. R., Peirano, P., Birch, D. G., & Birch, E. E. (2001). Essential fatty acids in visual and brain development. *Lipids, 36*, 885–895.

Yan, Y., Peng, H., Wang, P., Wang, H., & Dong, M. (2016). Increased expression of fatty acid binding protein 4 in preeclamptic Placenta and its relevance to preeclampsia. *Placenta, 39*, 94–100.

Yang, S. P., Morita, I., & Murota, S. I. (1998). Eicosapentaenoic acid attenuates vascular endothelial growth factor-induced proliferation via inhibiting Flk-1 receptor expression in bovine carotid artery endothelial cells. *Journal of Cellular Physiology, 176*, 342–349.

Yuan, Y. M., Fang, S. H., Qian, X. D., Liu, L. Y., Xu, L. H., Shi, W. Z., Zhang, L. H., Lu, Y. B., Zhang, W. P., & Wei, E. Q. (2009). Leukotriene D4 stimulates the migration but not proliferation of endothelial cells mediated by the cysteinyl leukotriene cyslt(1) receptor via the extracellular signal-regulated kinase pathway. *Journal of Pharmacological Sciences, 109*, 285–292.

Chapter 5
B Vitamins and Their Role on Trophoblast Growth and Development

Contents

5.1 Introduction

Folic acid is critical for growing infant particularly relevant to preterm and low birth weight in addition to neural tube defects (NTDs) and should be given as supplement for low birth weight (Orzalesi and Colarizi 1982). Folic acid is also known as vitamin B_9 and acts as a coenzyme in the biosynthesis of purine and pyrimidine precursors of nucleic acids, which are critically important for DNA synthesis of the growing fetus during pregnancy. Its natural form is known as folate, metabolized differently than its synthetic oxidized form of folic acid. Pregnant women are at risk for folate insufficiency because for its increased demand for rapid fetal growth, placental development, and enlargement of the uterus. Inadequate folate status may cause fetal malformations, impaired fetal growth, preterm delivery, and maternal anemia. Maternal folic acid deficiency during pregnancy has been associated with neural tube defects (NTDs), miscarriage, placental abruption, IUGR, and PE. Folic acid has the potential to affect the closure of the neural tube

A.K. Duttaroy, S. Basak, *Early Nutrition and Lifestyle Factors*,
DOI 10.1007/978-3-319-38804-5_5

and several epigenetic mechanisms within the placenta and the fetus. As fetal growth is totally dependent on maternal folic acid reserve, therefore, pregnant women are advised to receive folic acid supplement starting from periconceptional period. It is possible that folic acid plays other growth-promoting activities in early development such as implantation and development of the placenta and in improving endothelial function. Due to these indispensable functions, folic acid supplements are considered at the late first trimester or early second trimester (Fekete et al. 2010). Folic acid is increasingly required for rapid growth and cell proliferation during the initial phases of pregnancy. Folic acid deficiency is critical in development stages that determines long-term disease outcome. The chapter emphasizes the functional aspect of folic acid and its implication in human development during early placentation. This review discusses well-designed intervention trials to provide a latest update on the output of the efficacy of folic acid supplementation in mitigating and preventing disease etiology. The information is also obtained from observational studies where a large number of samples are associated. The priority is given to human clinical studies; if not available, animal data are reviewed; and in vitro studies are included to understand its mechanism on cellular function. Evidences are now accumulating that periconceptional folic acid supplement starting before 12 weeks of conception has several benefits on the pregnancy outcome through improved feto-placental growth and development activities (Cawley et al. 2016b) In this review, the impact of folic acid supplements on placental growth and development activities is discussed with special emphasis on their effects in early pregnancy.

5.2 Folic Acid Regulates Placental Growth Activities and Trophoblast Invasion

Folic acid has many physiological functions, including cell proliferation, DNA replication, and antioxidant protection as evidenced in different cellular systems. It has been implicated in the regulation of trophoblast invasion and placental development in normal early human pregnancy. Trophoblast is sensitive to many insults that may include poor nutrition during its differentiation or epithelization, a condition which can interfere with folic acid cycle. Interference of DNA synthesis by altered folic acid cycle could be associated with miscarriage. Increasing level of exogenous folic acid in vitro increases rates of proliferation, growth, and function of the extravillous HTR8/SVneo trophoblast cells (Ahmed et al. 2016). Direct effect of folic acid on trophoblast cells demonstrates that its presence can boost uterine invasion like function of the cellular trophoblast in vitro (Williams et al. 2011). Folic acid improves not only the factors associated with invasion of extravillous trophoblast (EVT) cells but also increases cellular proliferation and decreases apoptosis of the first-trimester placental explant. In addition, folic acid increases vascular density of the first-trimester placental culture, as well as the

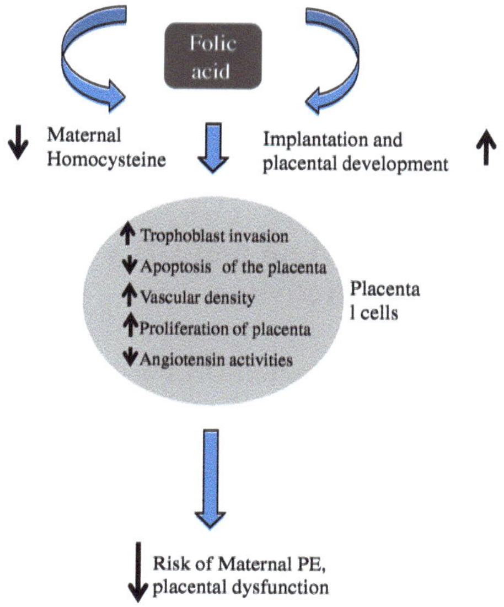

Fig. 5.1 Possible roles of folic acid in minimizing the risk of placental abnormalities

secretion of matrix metalloproteinases. The treatment with the antifolate drug methotrexate in ectopic pregnancy has been shown to reduce placental growth and EVT invasion due to reduced trophoblast proliferation. Folic acid increases angiogenesis in placental explant culture. Placental angiogenesis is crucial for the appropriate development of the blood flow and circulation of the feto-placental unit. So far, in vitro study with first-trimester placenta showed that folic acid may play a positive role in early stages of development by improving trophoblast invasion and angiogenesis processes. However, further studies are required with first-trimester trophoblast cells to evaluate its effect on these processes. Nevertheless, these studies support a notion on periconceptional folic acid supplementation to target specific pregnancy abnormalities where trophoblast growth and invasion are compromised. There are other studies where folic acid has been shown to regulate angiogenic growth factors. For example, folic acid improves expression of VEGF and normalizes plasma homocysteine level in angiotensin II-induced mice. Moreover, folic acid partially mitigates blood pressure parameters in this model by increasing vascular density and renal cortical blood flow and by decreasing the expression of anti-angiogenic factor (Pushpakumar et al. 2013). Deficiency of folic acid when combined with environmental and lifestyle factors such alcohol, lithium, or hyperhomocysteinemia, in the first 2–3 weeks of gestation, can result in severe congenital heart defects (CHDs). This vitamin can regulate Wnt signaling system by suppressing Wnt-mediated gene expression that result in a delay of cardiomyogenesis and CHD (Linask 2013). Preventive activity of folic acid on dexamethasone (Dex)-induced feto-placental growth restriction has been shown recently in mice. Folic acid protects Dex-induced growth restriction parameters in the placenta by elevating the expression of VEGF and PIGF activities possibly by modulating methylation of VEGF promoters (Zhang et al. 2015). Based on the available in vivo and in vitro data, it seems that folic acid improves several underlying steps of placentation and may impart protective effects in preventing placental dysfunction and maternal risk of pathological pregnancy (Fig. 5.1)

5.3 Folic Acid and DHA on Placental Growth and Activities

A combined effect of DHA and folic acid precursor, 5-methylenetetrahydrofolate (5-MTHF), is studied to evaluate its effects on the proliferation and apoptosis processes in the development of the placenta using a randomized double-blind placebo control of pregnant women ($n = 55$) who are supplemented with fish oil from the second trimester. The proliferation cell nuclear antigen (PCNA) level in the fish oil + 5-MTHF-treated group is significantly higher than that of the placebo group, suggesting that defined combination of DHA and 5-MTHF upregulates the placental proliferation process. The enhanced proliferation is found only in villous cytotrophoblasts, not with endothelial and other stromal cells (Klingler et al. 2006). Combination of folate and long-chain polyunsaturated fatty acid supplements during pregnancy possibly improves child's attention by promoting brain development of the fetus. This has been reported from NUHEAL (Nutraceuticals for a Healthy Life) study where children ($n = 136$) were born to mothers who had prenatal supplementation of fish oil and 5-MTHF. A positive long-term effect of this supplementation seems to improve the decision-making process of the children (8.5 years) by increased activation of the midcingulate cortex of the brain (Catena et al. 2016).

5.4 Maternal Folic Acid as a Determinant of Fetal Growth and Outcome

The relationship between body mass index (BMI) and serum folate concentrations in pregnant women in mid- ($n = 802$) and late ($n = 660$) pregnancy demonstrates a significant negative correlation (Kim et al. 2012). In order to investigate whether alteration in the folate metabolism is one of the causative factors in the pathogenesis of PE, folic acid metabolism markers are analyzed in placenta samples derived from PE ($n = 28$) and healthy ($n = 40$) women. Altered expression of cystathionine-beta-synthase (CBS) and accumulation of cysteine in placental explants when cultured in the presence of elevated homocysteine suggest an alteration in placental folic acid-related metabolic activity in developing preeclampsia (Mislanova et al. 2011). A prospective case-control clinical pilot study conducts a comparative analysis of maternal and fetal serum levels of homocysteine and folic acid and placental tissue levels of homocysteine and their association with the severity of PE ($n = 26$) and healthy ($n = 26$) subjects. Homocysteine levels are found to be higher in both maternal and fetal serum of the severe PE group compared to mild PE and control groups. However, elevated serum homocysteine levels are not associated with the deficiency of folic acid (Acilmis et al. 2011). The association between homocysteine, folic acid, and vitamin B_{12} concludes that hyperhomocysteinemia, deficiency of folic acid and vitamin B_{12}, and high blood pressure are common with PE ($n = 50$) as compared to healthy subjects ($n = 50$) (Mujawar et al. 2011). Data on Indian birth cohort reports an association between maternal homocysteine, folic acid concentration, and birth size with a follow-up risk factor for insulin resistance and glycemia in the offspring up to 13 years, suggesting a possible link between one-carbon metabolism and fetal programming of diabetes (Krishnaveni et al. 2014). A large birth cohort of population-based study from the Netherlands on pregnant women ($n = 5805$) concludes that higher homocysteine and low folic acid in early pregnancy are associated with lower placental weight and birth weight and higher risk of pregnancy (Bergen et al. 2012). An alteration in sperm epigenome due to folic acid deficiency is associated with negative pregnancy outcomes of the mice suggesting that even low paternal dietary folic acid can contribute to the function of the folic acid deficiency (Lambrot et al. 2013).

5.5 Transport and Metabolism of Folic Acid Across the Placenta

Folic acid is increasingly required for rapid growth and cell proliferation during the initial phases of pregnancy. Poor folate level during pregnancy can lead to elevated plasma level of homocysteine (Hcy) which is an independent risk factor for pregnancy-associated complications. Thus, placental metabolism of homocysteine is a key determinant for the feto-placental outcome. Expression levels of the genes encoding Hcy metabolism such as methionine synthase and 5,10-methylenetetrahydrofolate reductase are similar between first-trimester and term placenta as compared to the liver (positive control), while cystathionine-beta-synthase expression is lower in the liver as compared to the placenta from both gestations. This data indicates that re-methylation of Hcy by utilizing methyl donor from 5-methylenetetrahydrofolate is largely dependent on maternal folate availability. Folate transporters are localized and expressed in the first-trimester placenta since folic acid receptor alpha (FR-alpha) is detected in microvillous plasma membrane (MVM) of the syncytiotrophoblast in both the first-trimester and term placentas. A significant reduction in the level of folic acid carrier at term suggests its utilization during gestation. Folic acid transporter system is established early in pregnancy, which provides sufficient folate for utilization in placental homocysteine metabolism (Solanky et al. 2010). The mechanisms of membrane transport of folic acid into the cells and across epithelia are not clearly understood (Zhao et al. 2011). The folic acid metabolic enzymes, methionine synthase (MTR) and 5,10-methylenetetrahydrofolate reductase (MTHFR), are localized in the placental tissue. MTR is expressed in the villous syncytiotrophoblast involved in the metabolism of the homocysteine. MTHFR, which is expressed in the extravillous trophoblast, may be associated with extravillous trophoblast invasion. Differential expression of these two enzymes suggests specific function in folate metabolism in a time- and tissue-dependent manner (Shin et al. 2014). Folate deficiency has been associated with single nucleotide polymorphisms (SNPs) of folate metabolic genes such as methionine synthase (MTR) and methylenetetrahydrofolate dehydrogenase 1 (MTHFD1). Human placental choriocarcinoma (JEG-3) cells showed lower proliferation and invasion capacity when cultured in low folate condition (2 nM vs. 20 nM of folic acid). Silencing of the MTR gene and protein results in a decrease in the proliferation and an increase in the secretion of progesterone without affecting cell invasion. Silencing of MTHFD1 gene had no effects on such activities (Moussa et al. 2015). Placental uptake of ^{3}H-folic acid is pH dependent, increases significantly with decreasing extracellular pH. Whether maternal diabetes in gestation modulates folic acid supply to the placenta and fetus is not known. Folic acid uptake by syncytiotrophoblast is affected by gestational diabetes (Araújo et al. 2013). Serum folate level is altered in maternal blood and umbilical cord blood at the time of delivery in pregnancies with chronic and heavy alcohol exposure ($n = 23$) as compared to nondrinker ($n = 24$). In the alcohol-exposed pairs, the fetal/maternal serum folate ratio is ≤ 1.0 in more than 50 % of the subjects, indicating that the transport of folate from the mother to the fetus is affected upon heavy alcohol intake (Hutson et al. 2012). Maternal smoke does not impair folate transfer directly in early pregnancy (Jauniaux et al. 2007). The use of antifolic acid drug such as trimethoprim during pregnancy could be a reason for miscarriage. Trophoblasts are very sensitive to drugs that interfere with the folic acid cycle. Trimethoprim possibly inhibits DNA synthesis of trophoblast cells during the first trimester (Andersen et al. 2013).

5.6 Periconceptional Nutrition, Folic Acid Supplementation, and Pregnancy Outcome

Maternal store supplies the nutrients to the fetus via the placenta. Therefore, nutrient status during pregnancy determines the feto-placental growth and development. Several studies have already been done to examine the role of nutrients typically on the second or third trimester; by that time development and/organogenesis process have been completed. However, nutrient status of pregnant women just before conception and/or during the first trimester, when women are typically unaware of their pregnancy status, may influence pregnancy outcome by affecting critical development processes that happened in early pregnancy. Knowledge on periconceptional use of micronutrient is not uniform among the population across the nations, e.g., folic acid for the prevention of neural tube defects is less known, and periconceptional use of folic acid is lower in South Asians compared to Caucasian (Peake et al. 2013). In order to measure the impact of maternal nutrition on early placentation, the intervention data on nutritional supplement during periconceptional period is necessary. The practice of preconception and periconception supplementation of vitamin and minerals on parameters of pregnancy outcome relative to fetal growth and birth size has been compiled (Cetin et al. 2010). Based on the systematic review of the impact of maternal nutrition before and during early pregnancy (<12 weeks gestation) on maternal, neonatal, and child health, outcomes show that periconceptional (<12 weeks gestation) folic acid supplementation significantly reduced the risk of neural tube defects. The benefit of supplementation during maternal preconception in case of folic acid against the risk of NTD is well documented but could not have an effect in preventing stillbirth (Ramakrishnan et al. 2012).

Although observational studies suggest that preconceptional and periconceptional intake of vitamin and mineral supplements is associated with a reduced risk of delivering offspring who are low birth weight and/or small for gestational age (SGA) and preterm deliveries (PTD), however, more intervention and functional studies would be required to establish their benefit in those cases.

Controversy arises for the prenatal folic acid supplementation when exceeding the upper intake level in several countries. Conducting a study with excess folic acid is not possible with human. In such cases, in vitro study provides an initial basis for the future clinical trial. Recent study with increasing concentration of exogenous folic acid is tested for trophoblast cell growth and function in vitro. Excess folic acid (2000 ng/ml) increases the rates of proliferation, while too little (2 ng/ml) decreased cell viability and invasive capacity of the HTR8/SVneo cells in vitro (Ahmed et al. 2016). Periconceptional supplement with folic acid is reported to mediate several protective effects against the incidence of NTD. It is not known whether homocysteine toxicity or homocysteinylation induced by folate deficiency is mitigated by folic acid supplementation. Recent data with low-folate diet not only induces NTD in mice but also increases protein homocysteinylation and the subsequent generation of autoantibodies against homocysteinylated proteins. Homocysteinylation results in neo-self antigen formation under conditions of maternal folate deficiency, and the process is reversible with folic acid supplementation (Denny et al. 2016). Epigenetics influence of periconceptional folic acid supplementation could be a possible benefit. A large-scale genome-based approach is adopted to investigate the association between maternal plasma folate and epigenome-wide DNA methylation during pregnancy from 1988 newborn cases of two European cohorts. Four hundred and forty-three CpGs (320 genes) are significantly associated with maternal plasma folate levels during pregnancy which are related with birth defects, neurological defect, and many aspects of embryonic development apart from NTD (Joubert et al. 2016). This study marks distinct evidence on the possible role of maternal folate in developing baby's epigenome and the parameters of possible health outcome in the offspring.

Observational study from the pregnant women in their first trimester ($n = 587$) concludes that 96 % of the women started folic acid only after become pregnant in Ireland. Data indicates prepregnancy folic acid supplementation is not being implemented in Ireland against the current recommendation which may be due to lack of awareness or confidence on this recommendation. It seems more data on long-term effects of prepregnancy supplementation of folic acid on both safe use of mother and the offspring are possibly required to ensure implementation of the current recommendation among the population (Cawley et al. 2016a). The prevalence of periconceptional folic acid supplementation in France is evaluated by a cross-sectional population-based study ($n = 12{,}646$) using interview method after delivery. Only 14.8 % of women used folic acid before pregnancy, and the percentage is apparently linked with educational levels of women (Tort et al. 2013). A population-based case-control study investigated the association between periconceptional folic acid supplementation and uteroplacental vascular resistance, blood pressure, and the risks of gestational hypertension and preeclampsia in the pregnant women ($n = 5993$). Periconceptional folic acid supplementation is associated with lower uteroplacental vascular resistance and higher blood pressures during pregnancy but the effects are small and within physiologic ranges and are not associated with the risk of hypertensive pregnancy disorders (Timmermans et al. 2011). The periconceptional effect of folic acid supplement is also investigated in the context of methylation abilities of this vitamin. Periconceptional maternal folic acid supplement affects the methylation of the insulin-like growth factor 2 gene and is associated with epigenetic changes of the IGF-2 gene possibility due to intrauterine reprogramming. Thus, methylation of IGF-2 pathways is altered by periconceptional folic acid supplement (Steegers-Theunissen et al. 2009).

Periconceptional folic acid supplements reduce the risk of neural tube defects which is established. A large prospective cohort from Norway suggesting maternal consumption of supplements containing folic acid during the periconceptional period of 4 weeks before till 8 weeks after conception is associated with a reduction in the risk of the child having severe language delay at the age of 3 years (Roth et al. 2011). The trial started from week 17 of pregnancy and continued to follow supplementation effects on neurodevelopment parameters by measuring language learning of the 3-years children after birth. This is a unique study to postulate a hitherto unknown implication of periconceptional folic acid supplementation on neurodevelopmental function. This study would certainly influence to consider folic acid supplementation window before pregnancy in many countries. The effect of folic acid supplementation in preventing the risk factors of pregnancy outcome is summarized based on available data from clinical studies (Fig. 5.2).

Since conception to birth, an offspring is completely dependent on the nutrition of the mother via the placenta. The priority problems of the intervention would be a must to include supplementation with iron-folic acid or multiple micronutrients (Mason et al. 2014). In Europe, recommendation for folic acid supplements varies country wise and also the rate of NTD. Few guidelines recommend starting folic acid at least 4 weeks preconceptionally, but no recommendation of commencing folic acid supplement at least 3 months preconceptionally although recent data are

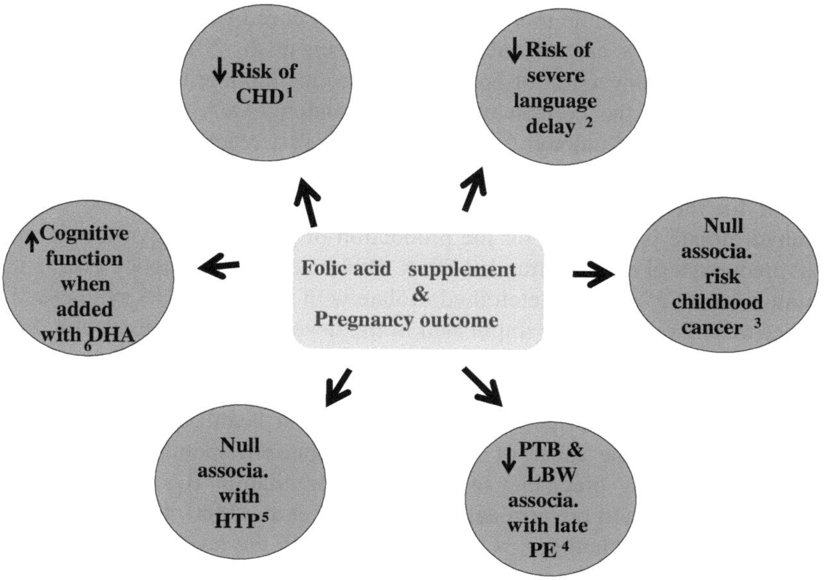

Fig. 5.2 Effects of folic acid supplementation on the risk factors of pregnancy outcome. *CHD* congenital heart disease, *PTB* preterm birth, *LBW* low birth weight, *PE* preeclampsia, *HTP* hypertensive pregnancy, *DHA* docosahexaenoic acid. *1* Czeizel et al. (2015), *2* Roth et al. (2011), *3* Mortensen et al. (2016), *4* Banhidy et al. (2011), *5* Timmermans et al. (2011), *6* Catena et al. (2016)

favoring for its beneficial effects when taken as periconceptional (Cawley et al. 2016b). A large variation of folic acid recommendation among the European nations may be due to fear of adverse effects of periconceptional folic acid supplement, lack of long-term data on possible health effects of the offspring, lack of awareness on the potential benefit of the supplement, or other unknown ethical factors. Therefore, more studies are recently being conducted to measure the adverse effects in any of folic acid supplementation during pregnancy (Mortensen et al. 2016; Czeizel et al. 2015; Cueto et al. 2016). However, a recent rise of NTD in Ireland raises a serious question to have a consensus document among these nations. Many urge a common guideline that can be followed in European Union.

5.7 Vitamin B_{12} Nutrition During Gestation

Vitamin B_{12} acts as a cofactor in many metabolic processes, and its deficiency is associated with megaloblastic anemia and several neurological disorders. Vitamin B_{12} (vitB_{12}) is another key micronutrient during pregnancy which together with folate regulates one-carbon metabolism pathway. Unlike prokaryotes, humans and other mammals are unable to synthesize B_{12}. The core of B_{12} consists of a corrin ring that encircles a central cobalt ion (Nielsen et al. 2012). B_{12} serves as a coenzyme in the conversion of homocysteine to methionine by cytosolic methionine synthase and the conversion of methylmalonyl-CoA to succinyl-CoA by mitochondrial methylmalonyl-CoA mutase. The first reaction is linked to folate metabolism because the methyl group is transferred to homocysteine which is required for the conversion of 5′-methylenetetrahydrofolate to tetrahydrofolate. Tetrahydrofolate is essential for the production of purines and pyrimidine, nucleotide precursors of double-stranded DNA. Prolonged B_{12} deficiency results in the accumulation of 5′-methylenetetrahydrofolate with impaired DNA synthesis. Deficiencies of vitB_{12} have been implicated in the developmental defects and delayed maturity and the development of neural tube defects, birth defect, adiposity, coronary artery disease, and autism spectrum disorders and other neurodevelopmental disorders. In terms of maternal health, clinical vitamin B_{12} deficiency may be the cause of infertility or recurrent spontaneous abortion or miscarriage. Starting pregnancy with inadequate vitamin B_{12} may increase the risk of birth defect and preterm delivery (Molloy et al. 2008). However, placental disease, such as preeclampsia, is not associated with the level of vitamin B_{12} and elevated serum level of homocysteine (Acilmis et al. 2011). Recent data reported that the prevalence of lower maternal B_{12} level and high folate status is common with Indian population. Being a vegetarian country, like India, it is not unusual to have a vitamin B_{12} deficiency in the population. Vitamin B_{12} deficiency, along with low-protein and high-carbohydrate intake of a vegetarian diet, may have a cascading deficiency effects to the pregnant women of the developing countries.

5.8 Vitamin B_{12} Restriction Affects Growth and Metabolism of the Nutrients

Evidence is accumulating that maternal B_{12} status influences fetal growth and development. Low maternal vitamin B_{12} status and protein intake are associated with increased risk of neural tube defect, low lean mass and excess adiposity, increased insulin resistance, impaired neurodevelopment, and altered risk of cancer in the offspring (Rush et al. 2014). Maternal vitamin B_{12} and folic acid restriction altered lipid metabolism in the rat offspring and increased visceral adiposity (Kumar et al. 2013). Offspring born to vitamin B_{12}-restricted dams had low birth weight than the control one. Vitamin B_{12} deficiency is found to be linked with growth and metabolism pathways. Animal data from maternal vitamin B_{12} restriction suggest alteration in the expression of proteins associated with two vital lipid metabolism pathways that included fatty acid oxidation and PPAR gamma pathways. Downregulation of beta oxidation possibly indicates that vitamin B_{12} deficiency regulates systemic energy partition and favors lipid accumulation processes that increased adiposity of these animals under vitB_{12} restrictive condition (Ahmad et al. 2013). PPAR gamma is the master switch for lipid metabolism and acts as a natural ligand for long-chain fatty acid; therefore, alteration of LCFAs metabolism is likely in case of vitamin B_{12} deficiency. VitB_{12} levels of the placenta are associated with trophoblast invasion markers such as matrixmetalloproteinases (MMPs). Placental levels of matrix metalloproteinase in term ($n = 75$) and preterm ($n = 73$) women are positively associated with folate and vitB_{12} level in preterm birth. Since MMPs actively participate in the remodeling of placental trophoblast invasion during placentation, thus, these vitamins (folic acid and B_{12}) may play a role in the process of placentation by regulating the activities of these enzymes (Sundrani et al. 2013).

5.9 Vitamin B_{12} Deficiency Can Be Predicted from Urine in Pregnancy?

Vitamin B_{12} is essential for normal placentation and fetal growth and development. This vitamin is ingested through a gastric intrinsic factor-mediated uptake in the ileal enterocytes by the mother during pregnancy. After absorption, B_{12} bound to transcobalamin (TC). The circulatory form of TC is partly in free form (apo-TC) and remaining with B_{12} (holo-TC). The holo-TC form of B_{12} is required for the receptor-mediated cellular uptake of B_{12}. The holo-TC binding receptor named CD320 is recently purified from the placenta. The CD320 receptor belongs to low-density lipoprotein receptor. The soluble extracellular domain (sCD320) is heavily glycosylated with MW58kDa. The placenta expresses abundantly with this CD320 receptor, and increasing serum level of sCD320 up to 35 weeks possibly contributes from the placental side. The function, mechanism, and regulation of the receptor sCD320 and its release in the placenta are largely unknown. The soluble form of sCD320 is increased from 125 (87–839) pmol/L (week 15) to reach a peak value of 199 (72–672) pmol/L (week 35) and subsequently decreased to its baseline just before birth (week 40). The urinary level of sCD320 is around twofold higher than serum sCD320 value although both serum and urine showed the similar elution pattern upon size exclusion chromatography (Abuyaman et al. 2013). This further raises the possibility to estimate the vitamin B_{12} level from the urine. Moreover, the level of this receptor in the placenta in the pathological pregnancy needs to be studied. Measuring the cellular uptake of vitamin B_{12} is an important process to ascertain their availability for fetal growth and development. Binding of vitamin B_{12} to transcobalamin (TC) makes them available to transport from the maternal serum to the placenta by the receptor CD320. The soluble form of CD320 (sCD320) is elevated in pregnancy and an important indicator of vitamin B_{12} availability. The secretion of this receptor and holo-TC is not known in miscarriage and preeclampsia. Based on the available data, the impact of maternal vitamin B_{12} deficiency during pregnancy on various risk factors of the offspring later in life is tabulated (Table 5.1).

Table 5.1 Impact of maternal vitamin B_{12} deficiency on the risk of fetal health and disease

Deficiency of B_{12}	Risk factors	References
Maternal vitamin B_{12}	↑ Birth defects ↑ Preterm delivery	Molloy et al. (2008)
Low maternal vitamin B_{12} and low protein intake	↑ NTD ↑Cancer Impaired neurodevelopment	Rush et al. (2014)
Low maternal vitamin B_{12} and high folate	↑ Adiposity ↑ Type 2 diabetes	Yajnik et al. (2008)
Prolonged vitamin B_{12} deficiency	Impaired DNA synthesis due to accumulation of 5-MTHF	Nielsen et al. (2012)
Maternal soluble receptor sCD320 (urine and serum)	Vitamin B_{12} availability	Abuyaman et al. (2013)

5.10 Vitamin B_{12} Supplementation Corrects Deficiency During Pregnancy

Vitamin B_{12} deficiency is more pronounced in resource-limited countries. A prospective Pune maternal study with pregnant women ($n = 700$) predicted a positive association between maternal vitamin B_{12}, folate, and homocysteine status in pregnancy and offspring adiposity and insulin resistance later at 6 years. Two-thirds of mothers had low vitamin B_{12} level (<150 pmol/l). Authors suggest low maternal vitamin B_{12} and high folate status may contribute to the epidemic of adiposity and type 2 diabetes in India (Yajnik et al. 2008). Frequencies of vitamin B_{12} and folate deficiencies in pregnant women and neonates are also reported high in the Aegean region of Turkey, where the Mediterranean diet (mainly fresh fruits and vegetables) is adopted (Balci et al. 2014). Analysis of small samples of expectant mothers ($n = 72$) and cord blood of babies ($n = 72$) showed 70.8 % of the mothers and 83.9 % of the babies are vitamin B_{12} deficient (<200 pg/mL) without any folate deficiency. In contrast to these studies, prospective birth cohort from the Netherlands did not observe any association of vitamin B_{12} level with placental weight and birth weight of the pregnant women ($n = 5805$), although higher homocysteine and lower folate levels are associated (Bergen et al. 2012). All these data suggest that prenatal screening of all expectant mothers, prenatal supplementation of vitamin B_{12}, and an increase in animal-source food intake may be adopted to improve expectant mother's vitamin B_{12} level in these populations.

Recent human clinical trial on vitamin B_{12} supplementation during pregnancy demonstrated a positive boost for both the mother and infant vitB_{12} status. The randomized, placebo-controlled clinical trial included pregnant women <14 weeks of gestation in Bangalore, India (Duggan et al. 2014). They received daily oral supplementation with ($n = 183$) vitamin B_{12} (50 μm) or placebo ($n = 183$) through 6 weeks postpartum. All women administered iron and folic acid supplements throughout pregnancy. The vitamin B_{12} supplementation significantly improved plasma vitamin B_{12} concentrations of the mother at both the second (median vitamin B_{12}, 216 vs. 111 pmol/L, $P < 0.001$) and third (median, 184 vs. 105 pmol/L, $P < 0.001$) trimesters compared to placebo. At 6-week postpartum, median breast milk vitamin B_{12} level is 136 pmol/L in vitamin B_{12}-supplemented group compared to 87 pmol/L ($P < 0.0005$) in placebo. Infant's plasma methylmalonic acid and homocysteine concentrations are significantly lower in the vitamin B_{12} group. This would be interesting to follow up the supplemented infant at least with some of the parameters related to neurobehavioral and metabolic genotypes. The study possibly first time showed oral supplementation of vitamin B_{12} improved B_{12} level significantly in mother, breast milk, and babies. However, further study with increased number of samples in a randomized case control design involving vegetarian pregnant women is required. Nevertheless, this study raised a possibility to supplement this vitamin to overcome B_{12} deficiency in the deficient pregnant women.

5.11 Multivitamin or Only Vitamin Supplementation for Better Pregnancy Outcome

Early development of the placenta and fetus requires all the micronutrients for growth and survival especially in development stages spanning from conception, implantation, invasion, and proliferation that leads to functional pregnancy. However, optimal doses, types, and combinations of micronutrient requirement on the category of nutritionally stratified pregnant women are not known. Previous studies on periconceptional use of multivitamins in well-nourished women indicate beneficial effects against the risk factors of preeclampsia (Catov et al. 2009; Bodnar et al. 2006), intrauterine growth restriction, and preterm birth (Catov et al. 2007, 2011) possibly by affecting the early development and function of the placenta. Several clinical trials reported the efficacy of folic acid supplement on preventing the risk of pregnancy-related disorders to improve birth outcome (Nohr et al. 2014; Wen et al. 2008, 2013; Li et al. 2013; Timmermans et al. 2011; Roth et al. 2011; Banhidy et al. 2011; Charles et al. 2005; Hernandez-Diaz et al. 2002). Based on available evidences, periconceptional folic acid supplement has been implemented since the 1990s in many countries with high-dose folic acid (4.0–5.0 mg) recommended for high-risk women and low-dose folic acid (0.4–1.1 mg) for low-risk women in the prevention of NTDs (Wen et al. 2013).

Recent published clinical data on possible adverse outcome of multivitamin supplementation (MVS) in the industrialized developed country warrants further debate on its blanket recommendation. The effect of maternal multivitamins with or without folic acid on the prevalence of miscarriage is estimated in ($n = 35{,}914$) pregnant women derived from the Danish National Birth Cohort that included the number of weeks of supplement used by pregnant women during a 12-week periconceptional period (Nohr et al. 2014). The consumption of folate-containing multivitamin preparations in the periconceptional period was associated with a modest, but increased, risk of miscarriage in the early weeks of pregnancy, when compared with women who took no supplements. More specifically, regular users of multivitamins (4–6 weeks of 6) before conception had more early losses compared to late user. In contrast, folic acid-only use was not associated with fetal death. The reason for the fetal death after multivitamin intake is not clear especially when these are taken in preconception period. Large dose of vitamin toxicity with vitamin A is reported in gestation but their level present in the multivitamin ruled out the possibility of the toxicity. Authors pointed that apart from a decreased risk of low birth weight in multivitamin users, there is no convincing evidence from randomized trials of a beneficial effect on pregnancy outcomes. The major strength of the study lies in the large size of the cohort and the benefit of the detailed dataset afforded by the extraordinary depth and breadth of the Danish National Birth Cohort. In the commentary of the article, Poston (Poston 2014) made a well-balanced conclusive remark that obviously warrant an additional well-designed randomized control trial (RCTs) which compare pregnancy outcomes, particularly early fetal loss, in women taking multivitamins in the periconceptional period and in pregnancy with those taking folate alone, including measurement of folate status. Considering the unknown risk associated with the use of multivitamin in periconceptional period of healthy pregnant women, there could be fewer possibilities on the general recommendation of multivitamin supplementation immediately without having additional studies on this aspect. Many fresh doubts will be raised on MVS, whether MVS works better on targeted approach of the stratified nutritional background possibly in a combination of one or two. Can be vitamin requirement window for pregnant women universally recommended?

5.12 Summary

The role of folic acid on prevention of neural tube defects is well established. Emerging reports are suggesting folic acid can be protective against other adverse development. Available data collected from many studies suggest that periconceptional supplementation of folic acid could be the most effective way in preventing the risk factor associated with some of the pregnancy-associated complications. Overall data indicate that folic acid is indispensible for the developing fetus, and their deficiency may affect fertility, embryogenesis, and placentation. The prophylactic use of folic acid may be useful in preventing adverse pregnancy outcomes. However, further research is necessary to optimize periconceptional micronutrient requirements. Recent evidence of human supplementation is conclusive in favor of periconceptional folate intake in order to prevent neural tube defects. High magnitude of vitamin B_{12} deficiency is evidenced from India. The long-term effects of this deficiency on the fetus genotypes due to programming are not known. Characterization of the deficiency and their implication on various disease and development pathways would be required to understand its causality. Oral supplementation of vitamin B_{12} to the pregnant women showed a positive response against the deficiency. Further study is needed to evaluate its impact on neurodevelopment and other functional outcomes of the pregnancy. Overall data suggest multivitamin supplementation may need more selective approach for the cases based on well-designed data. The combination of dual vitamin-based targeted approach proved to be more effective than multivitamin supplement. Supplementation alone would be ineffective unless efforts are made to increase awareness of a healthy diet starting well before the pregnancy to mitigate the deficiency.

References

Abuyaman, O., Andreasen, B. H., Kronborg, C., Vittinghus, E., & Nexo, E. (2013). The soluble receptor for vitamin B12 uptake (sCD320) increases during pregnancy and occurs in higher concentration in urine than in serum. *PLoS One, 8*.

Acilmis, Y. G., Dikensoy, E., Kutlar, A. I., Balat, O., Cebesoy, F. B., Ozturk, E., Cicek, H., & Pence, S. (2011). Homocysteine, folic acid and vitamin B12 levels in maternal and umbilical cord plasma and homocysteine levels in placenta in pregnant women with pre-eclampsia. *Journal of Obstetrics and Gynaecology Research, 37*, 45–50.

Ahmad, S., Kumar, K. A., Basak, T., Bhardwaj, G., Yadav, D. K., Lalitha, A., Chandak, G. R., Raghunath, M., & Sengupta, S. (2013). PPAR signaling pathway is a key modulator of liver proteome in pups born to vitamin B12 deficient rats. *Journal of Proteomics, 91C*, 297–308.

Ahmed, T., Fellus, I., Gaudet, J., Macfarlane, A. J., Fontaine-Bisson, B., & Bainbridge, S. A. (2016). Effect of folic acid on human trophoblast health and function in vitro. *Placenta, 37*, 7–15.

Andersen, J. T., Petersen, M., Jimenez-Solem, E., Broedbaek, K., Andersen, E. W., Andersen, N. L., Afzal, S., Torp-Pedersen, C., Keiding, N., & Poulsen, H. E. (2013). Trimethoprim use in early pregnancy and the risk of miscarriage: A register-based nationwide cohort study. *Epidemiology and Infection, 141*, 1749–1755.

Araújo, J. R., Correia-Branco, A., Moreira, L., Ramalho, C., Martel, F., & Keating, E. (2013). Folic acid uptake by the human syncytiotrophoblast is affected by gestational diabetes, hyperleptinemia, and TNF-α. *Pediatric Research, 73*, 388–394.

Balci, Y. I., Ergin, A., Karabulut, A., Polat, A., Doğan, M., & Küçüktaşci, K. (2014). Serum Vitamin B12 and Folate Concentrations and the Effect of the Mediterranean Diet on Vulnerable Populations. *Pediatric Hematology and Oncology, 31*(1), 62–7.

Banhidy, F., Dakhlaoui, A., Dudas, I., & Czeizel, A. E. (2011). Birth outcomes of newborns after folic acid supplementation in pregnant women with early and late pre-eclampsia: A population-based study. *Advances in Preventive Medicine, 2011*, 127369.

Bergen, N. E., Jaddoe, V. W. V., Timmermans, S., Hofman, A., Lindemans, J., Russcher, H., Raat, H., Steegers-Theunissen, R. P. M., & Steegers, E. A. P. (2012). Homocysteine and folate concentrations in early pregnancy and the risk of adverse pregnancy outcomes: the Generation R Study. *BJOG, 119*, 739–751.

Bodnar, L. M., Tang, G., Ness, R. B., Harger, G., & Roberts, J. M. (2006). Periconceptional multivitamin use reduces the risk of preeclampsia. *American Journal of Epidemiology, 164*, 470–477.

Catena, A., Munoz-Machicao, J. A., & Torres-Espinola, F. J. (2016). Folate and long-chain polyunsaturated fatty acid supplementation during pregnancy has long-term effects on the attention system of 8.5-y-old offspring: A randomized controlled trial. *American Journal of Clinical Nutrition, 103*, 115–27.

Catov, J. M., Bodnar, L. M., Ness, R. B., Markovic, N., & Roberts, J. M. (2007). Association of periconceptional multivitamin use and risk of preterm or small-for-gestational-age births. *American Journal of Epidemiology, 166*, 296–303.

Catov, J. M., Bodnar, L. M., Olsen, J., Olsen, S., & Nohr, E. A. (2011). Periconceptional multivitamin use and risk of preterm or small-for-gestational-age births in the Danish National Birth Cohort. *American Journal of Clinical Nutrition, 94*, 906–912.

Catov, J. M., Nohr, E. A., Bodnar, L. M., Knudson, V. K., Olsen, S. F., & Olsen, J. (2009). Association of periconceptional multivitamin use with reduced risk of preeclampsia among normal-weight women in the Danish National Birth Cohort. *American Journal of Epidemiology, 169*, 1304–1311.

Cawley, S., Mullaney, L., Mckeating, A., Farren, M., Mccartney, D., & Turner, M. J. (2016a). An analysis of folic acid supplementation in women presenting for antenatal care. *Journal of Public Health (Oxford, England), 38*, 122–129.

Cawley, S., Mullaney, L., Mckeating, A., Farren, M., Mccartney, D., & Turner, M. J. (2016b). A review of European guidelines on periconceptional folic acid supplementation. *Nature Communications, 70*, 143–154.

Cetin, I., Berti, C., & Calabrese, S. (2010). Role of micronutrients in the periconceptional period. *Human Reproduction Update, 16*, 80–95.

Charles, D. H., Ness, A. R., Campbell, D., Smith, G. D., Whitley, E., & Hall, M. H. (2005). Folic acid supplements in pregnancy and birth outcome: Re-analysis of a large randomised controlled trial and update of Cochrane review. *Paediatric and Perinatal Epidemiology, 19*, 112–124.

Cueto, H. T., Riis, A. H., Hatch, E. E., Wise, L. A., Rothman, K. J., Sorensen, H. T., & Mikkelsen, E. M. (2016). Folic acid supplementation and fecundability: A Danish prospective cohort study. *European Journal of Clinical Nutrition, 70*, 66–71.

Czeizel, A. E., Vereczkey, A., & Szabo, I. (2015). Folic acid in pregnant women associated with reduced prevalence of severe congenital heart defects in their children: A national population-based case-control study. *European Journal of Obstetrics, Gynecology, and Reproductive Biology, 193*, 34–39.

Denny, K. J., Kelly, C. F., Kumar, V., Witham, K. L., Cabrera, R. M., Finnell, R. H., Taylor, S. M., Jeanes, A., & Woodruff, T. M. (2016). Autoantibodies against homocysteinylated protein in a mouse model of folate deficiency-induced neural tube defects. *Birth Defects Research Part A: Clinical and Molecular Teratology, 106*(3), 201–207.

Duggan, C., Srinivasan, K., Thomas, T., Samuel, T., Rajendran, R., Muthayya, S., Finkelstein, J. L., Lukose, A., Fawzi, W., Allen, L. H., Bosch, R. J., & Kurpad, A. V. (2014). Vitamin B-12 supplementation during pregnancy and early lactation increases maternal, breast milk, and infant measures of vitamin B-12 status. *Journal of Nutrition, 144*, 758–764.

Fekete, K., Berti, C., Cetin, I., Hermoso, M., Koletzko, B. V., & Decsi, T. (2010). Perinatal folate supply: Relevance in health outcome parameters. *Maternal and Child Nutrition, 6*(Suppl 2), 23–38.

Hernandez-Diaz, S., Werler, M. M., Louik, C., & Mitchell, A. A. (2002). Risk of gestational hypertension in relation to folic acid supplementation during pregnancy. *American Journal of Epidemiology, 156*, 806–812.

Hutson, J. R., Stade, B., Lehotay, D. C., Collier, C. P., & Kapur, B. M. (2012). Folic acid transport to the human fetus is decreased in pregnancies with chronic alcohol exposure. *PLoS One, 7*, e38057.

Jauniaux, E., Johns, J., Gulbis, B., Spasic-Boskovic, O., & Burton, G. J. (2007). Transfer of folic acid inside the first-trimester gestational sac and the effect of maternal smoking. *American Journal of Obstetrics and Gynecology, 197*, 58.e1–58.e6.

Joubert, B. R., den Dekker, H. T., Felix, J. F., Bohlin, J., Ligthart, S., Beckett, E., & Tiemeier, H. (2016). Maternal plasma folate impacts differential DNA methylation in an epigenome-wide meta-analysis of newborns. *Nature Communications, 7*, 10577.

Kim, H., Hwang, J. Y., Kim, K. N., Ha, E. H., Park, H., Ha, M., Lee, K. Y., Hong, Y. C., Tamura, T., & Chang, N. (2012). Relationship between body-mass index and serum folate concentrations in pregnant women. *European Journal of Clinical Nutrition, 66*, 136–138.

Klingler, M., Blaschitz, A., Campoy, C., Cano, A., Molloy, A. M., Scott, J. M., Dohr, G., Demmelmair, H., Koletzko, B., & Desoye, G. (2006). The effect of docosahexaenoic acid and folic acid supplementation on placental apoptosis and proliferation. *The British Journal of Nutrition, 96*, 182–190.

Krishnaveni, G. V., Veena, S. R., Karat, S. C., Yajnik, C. S., & Fall, C. H. D. (2014). Association between maternal folate concentrations during pregnancy and insulin resistance in Indian children. *Diabetologia, 57*(1), 110–21.

Kumar, K. A., Lalitha, A., Pavithra, D., Padmavathi, I. J. N., Ganeshan, M., Rao, K. R., Venu, L., Balakrishna, N., Shanker, N. H., Reddy, S. U., Chandak, G. R., Sengupta, S., & Raghunath, M. (2013). Maternal dietary folate and/or vitamin B12 restrictions alter body composition (adiposity) and lipid metabolism in Wistar rat offspring. *The Journal of Nutritional Biochemistry, 24*, 25–31.

Lambrot, R., Xu, C., Saint-Phar, S., Chountalos, G., Cohen, T., Paquet, M., Suderman, M., Hallett, M., & Kimmins, S. (2013). Low paternal dietary folate alters the mouse sperm epigenome and is associated with negative pregnancy outcomes. *Nature Communications, 4*, 2889.

Li, Z., Ye, R., Zhang, L., Li, H., Liu, J., & Ren, A. (2013). Folic acid supplementation during early pregnancy and the risk of gestational hypertension and preeclampsia. *Hypertension, 61*, 873–9.

Linask, K. K. (2013). The heart-placenta axis in the first month of pregnancy: Induction and prevention of cardiovascular birth defects. *Journal of Pregnancy, 2013*, 320413.

Mason, J. B., Shrimpton, R., Saldanha, L. S., Ramakrishnan, U., Victora, C. G., Girard, A. W., Mcfarland, D. A., & Martorell, R. (2014). The first 500 days of life: Policies to support maternal nutrition. *Global Health Action, 7*, 23623.

Mislanova, C., Martsenyuk, O., Huppertz, B., & Obolenskaya, M. (2011). Placental markers of folate-related metabolism in preeclampsia. *Reproduction, 142*, 467–476.

Molloy, A. M., Kirke, P. N., Brody, L. C., Scott, J. M., & Mills, J. L. (2008). Effects of folate and vitamin B12 deficiencies during pregnancy on fetal, infant, and child development. *Food and Nutrition Bulletin, 29*, S101–S111. discussion S112–115.

Mortensen, J. H., Oyen, N., Fomina, T., Melbye, M., Tretli, S., Vollset, S. E., & Bjorge, T. (2016). Supplemental folic acid in pregnancy and childhood cancer risk. *British Journal of Cancer, 114*, 71–75.

Moussa, C., Ross, N., Jolette, P., & Macfarlane, A. J. (2015). Altered folate metabolism modifies cell proliferation and progesterone secretion in human placental choriocarcinoma JEG-3 cells. *British Journal of Nutrition, 114*, 844–852.

Mujawar, S. A., Patil, V. W., & Daver, R. G. (2011). Study of serum homocysteine, folic Acid and vitamin b(12) in patients with preeclampsia. *Indian Journal of Clinical Biochemistry, 26*, 257–260.

Nielsen, M. J., Rasmussen, M. R., Andersen, C. B. F., Nexø, E., & Moestrup, S. K. (2012). Vitamin B12 transport from food to the body's cells—a sophisticated, multistep pathway. *Nature Reviews Gastroenterology and Hepatology, 9*, 345–354.

Nohr, E. A., Olsen, J., Bech, B. H., Bodnar, L. M., Olsen, S. F., & Catov, J. M. (2014). Periconceptional intake of vitamins and fetal death: A cohort study on multivitamins and folate. *International Journal of Epidemiology, 43*, 174–184.

Orzalesi, M., & Colarizi, P. (1982). Critical vitamins for low birthweight infants. *Acta Paediatrica Scandinavica - Supplement, 296*, 104–109.

Peake, J. N., Copp, A. J., & Shawe, J. (2013). Knowledge and periconceptional use of folic acid for the prevention of neural tube defects in ethnic communities in the United Kingdom: Systematic review and meta-analysis. *Birth Defects Research. Part A, Clinical and Molecular Teratology, 97*, 444–451.

Poston, L. (2014). Commentary: Multivitamins and early pregnancy loss. *International Journal of Epidemiology, 43*, 184–186.

Pushpakumar, S. B., Kundu, S., Metreveli, N., & Sen, U. (2013). Folic acid mitigates angiotensin-II-induced blood pressure and renal remodeling. *PLoS One, 8*, e83813.

Ramakrishnan, U., Grant, F., Goldenberg, T., Zongrone, A., & Martorell, R. (2012). Effect of women's nutrition before and during early pregnancy on maternal and infant outcomes: A systematic review. *Paediatric and Perinatal Epidemiology, 26*(Suppl 1), 285–301.

Roth, C., Magnus, P., Schjolberg, S., Stoltenberg, C., Suren, P., Mckeague, I. W., Davey Smith, G., Reichborn-Kjennerud, T., & Susser, E. (2011). Folic acid supplements in pregnancy and severe language delay in children. *JAMA, 306*, 1566–1573.

Rush, E. C., Katre, P., & Yajnik, C. S. (2014). Vitamin B12: One carbon metabolism, fetal growth and programming for chronic disease. *European Journal of Clinical Nutrition, 68*(1), 2–7.

Shin, J. A., Kim, Y. J., Park, H., Kim, H. K., & Lee, H. Y. (2014). Localization of folate metabolic enzymes, methionine synthase and 5,10-methylenetetrahydrofolate reductase in human placenta. *Gynecologic and Obstetric Investigation, 78*, 259–265.

Solanky, N., Requena Jimenez, A., D'Souza, S. W., Sibley, C. P., & Glazier, J. D. (2010). Expression of folate transporters in human placenta and implications for homocysteine metabolism. *Placenta, 31*, 134–143.

Steegers-Theunissen, R. P., Obermann-Borst, S. A., Kremer, D., Lindemans, J., Siebel, C., Steegers, E. A., Slagboom, P. E., & Heijmans, B. T. (2009). Periconceptional maternal folic acid use of 400 microg per day is related to increased methylation of the IGF2 gene in the very young child. *PLoS One, 4*, e7845.

Sundrani, D., Chavan-Gautam, P., Pisal, H., Mehendale, S., & Joshi, S. (2013). Matrix metalloproteinases-2, -3 and tissue inhibitors of metalloproteinases-1, -2 in placentas from preterm pregnancies and their association with one-carbon metabolites. *Reproduction (Cambridge, England), 145*, 401–410.

Timmermans, S., Jaddoe, V. W., Silva, L. M., Hofman, A., Raat, H., Steegers-Theunissen, R. P., & Steegers, E. A. (2011). Folic acid is positively associated with uteroplacental vascular resistance: the Generation R study. *Nutrition, Metabolism, and Cardiovascular Diseases, 21*, 54–61.

Tort, J., Lelong, N., Prunet, C., Khoshnood, B., & Blondel, B. (2013). Maternal and health care determinants of preconceptional use of folic acid supplementation in France: Results from the 2010 National Perinatal Survey. *BJOG, 120*, 1661–1667.

Wen, S. W., Champagne, J., Rennicks White, R., Coyle, D., Fraser, W., Smith, G., Fergusson, D., & Walker, M. C. (2013). Effect of folic acid supplementation in pregnancy on preeclampsia: The folic acid clinical trial study. *Journal of Pregnancy, 2013*, 294312.

Wen, S. W., Chen, X. K., Rodger, M., White, R. R., Yang, Q., Smith, G. N., Sigal, R. J., Perkins, S. L., & Walker, M. C. (2008). Folic acid supplementation in early second trimester and the risk of preeclampsia. *American Journal of Obstetrics and Gynecology, 198*, 45.e1–45.e7.

Williams, P. J., Bulmer, J. N., Innes, B. A., & Broughton Pipkin, F. (2011). Possible roles for folic acid in the regulation of trophoblast invasion and placental development in normal early human pregnancy. *Biology of Reproduction, 84*, 1148–1153.

Yajnik, C. S., Deshpande, S. S., Jackson, A. A., Refsum, H., Rao, S., Fisher, D. J., Bhat, D. S., Naik, S. S., Coyaji, K. J., Joglekar, C. V., Joshi, N., Lubree, H. G., Deshpande, V. U., Rege, S. S., & Fall, C. H. D. (2008). Vitamin B12 and folate concentrations during pregnancy and insulin resistance in the offspring: The Pune Maternal Nutrition Study. *Diabetologia, 51*, 29–38.

Zhang, A., Zhou, L. F., Xiang, X. L., Wang, K., Zhou, Q., & Duan, T. (2015). Folic acid attenuates dexamethasone-induced placental growth restriction. *European Review for Medical and Pharmacological Sciences, 19*, 1130–1140.

Zhao, R., Diop-Bove, N., Visentin, M., & Goldman, I. D. (2011). Mechanisms of membrane transport of folates into cells and across epithelia. *Annual Review of Nutrition, 31*, 177–201.

Chapter 6
Fat-Soluble and Antioxidant Vitamins and Minerals: Their Roles in Placentation

Contents

6.1 Introduction

Vitamins and minerals, often termed as micronutrients, serve essential roles in cellular metabolism, maintenance, and growth throughout life. Micronutrient deficiency is critical in development stages that determines long-term disease outcome. Micronutrients are also central components of many enzymes and transcription factors. The need for optimum amounts of key micronutrients at critical stages starting well before conception through ovulation, placentation, and fetus development has been documented recently. Both excesses and deficiencies of micronutrients can have long-term effects on many fetal tissues and organs due to subclinical deficiency of vitamins in the mother. Micronutrient imbalance of the developing fetus may not reflect any quantitative changes at the time of nutritional insult, but may have a permanent scar on genome imprint due to altered development circuit that manifests later in life (Hovdenak and Haram 2012). Unfortunately, supplementary correction of micronutrients later in gestation or during postnatal life cannot rescue or completely revert the detrimental effects of earlier micronutrient imbalance. In that context, periconceptional supplements and life course nutrition are being given a priority at this moment. The imbalance can affect pregnancy outcome through alterations in maternal and fetus metabolism, as a

A.K. Duttaroy, S. Basak, *Early Nutrition and Lifestyle Factors*,
DOI 10.1007/978-3-319-38804-5_6

Table 6.1 Requirement of vitamins and their role in feto-placental development [Adapted from Orzalesi and Colarizi (1982)]

Nature of vitamins	Fetus/maternal ratio (plasma)	Requirement of preconception women (per day)	Additional requirement in pregnancy (per day)
Lipid soluble vitamins transported predominantly by passive diffusion	Vitamin A = 1/1	4000 IU	1000 IU #
	Vitamin D = 0.8/1	400 IU	300 IU #
	Vitamin E = 0.3/1	12 IU	3 IU
	Vitamin K = 0.9/1	2 mg	1 mg
Water-soluble vitamins transported predominantly by active transport systems	Vitamin C = 2/1	45 mg	15 mg
	Vitamin B_1 = 2/1	1.1 mg	0.3 mg
	Vitamin B_2 = 4/1	1.4 mg	0.3 mg
	Vitamin B_6 = 3/1	2.0 mg	0.5 mg
	Vitamin B_{12} = 4/1	3.0 mcg	1.0 mcg
	Vitamin B_9 = 2/1	400 mcg	400 mcg

variable

consequence of their essential role in enzymes and transcription factors and through their involvement in signal transduction pathways that regulate development. Considering wide range of micronutrients which affects development, the number of developmental stages involved, diverse biological pathways associated, and types of tissue affected, it is challenging to pool all the micronutrients in this chapter. In this chapter, the risk associated with pregnancy due to excess or little of micronutrients in periconception and preconception periods that determine fetal health and development will be discussed. The chapter will discuss the functional aspect of fat-soluble vitamins and their implication in human development involving the placenta. Overall the chapter will emphasize the functional role of the fat-soluble vitamins and key minerals in feto-placental growth and development.

Apart from water-soluble vitamins such as folic acid, vitamin C, and vitamin B_{12}, fat-soluble vitamins, e.g., vitamin A, vitamin D, vitamin E, vitamin K, and the minerals zinc, iron, and copper are particularly important during pregnancy. The fat-soluble vitamins such as D, E, and K cross the placenta with difficulty. Therefore, vitamins such as D, E, K, and folic acid should be given as supplementation for low birth weight (Orzalesi and Colarizi 1982). The ratio of transfer of vitamins across the placenta varies between the fetus and mother due to the differences in solubility and transport mechanism of the vitamins, e.g., fat-soluble vitamins specially vitamin D, vitamin E, and vitamin K are accumulated less in the fetus in contrast to all water-soluble vitamins (Table 6.1). An adequate supply of lipid-soluble vitamins to the fetus largely depends on maternal level of vitamins, but transfer of water-soluble vitamins is independent of the levels in the maternal circulation.

6.2 Vitamin D and Its Role in Early Development

Vitamin D is known to have pleiotropic biological actions on several cellular and metabolic processes including on bone and calcium metabolism, immune functions, and responses to infections. Most of our vitamin D requirements (approx. 90 %) are met by sunlight through ultraviolet (UV-B) irradiation. Vitamin D_3 (cholecalciferol) is consumed in the diet from fish oil and fortified dairy products or synthesized in the skin from 7-dehydrocholesterol by sunlight. The amount of vitamin D produced by 7-dehydrocholesterol cycle depends on the intensity of UV irradiation, which is again influenced by several environmental factors such as clothing, pollution, season, latitude, and other factors in the conversion of 7-dehydrocholesterol to vitamin D (Christakos et al. 2013). Vitamin D is transported from the blood to the liver by vitamin D-binding protein (DBP). In the liver, vitamin D is hydroxylated to form 25-hydroxyvitamin D3 (25(OH)D3). The active form of vitamin D is 1,25-dihydroxyvitamin D(3) [1,25(OH)(2)D(3)]. The enzyme 1alpha-hydroxylase (1alpha-OHase) is the key to synthesize the active D3. This hydroxylase is expressed in several human tissues and acts as paracrine/autocrine activator of vitamin D. Protein for 1alpha-OHase is detected in trophoblast and more intensely in decidua. Expression (mRNA) of this enzyme is not similar throughout the gestation, e.g., mRNA expression of 1alpha-OHase is significantly higher in the first (mean, 9.1 ± 1.5 weeks) and second (mean, 14 ± 1.8 weeks) trimester when compared with the third trimester (mean, 39.3 ± 2.5 weeks) in both the placenta and in decidua. In parallel, expression of vitamin D receptor (1,25(OH)(2)D(3)) is also much higher in first-trimester decidua and does not change across gestation, emphasizing its role in early feto-placental life (Zehnder et al. 2002). This trend indicates an important role for 1,25(OH)2D3 in early pregnancy. The immunosuppressive effects of 1,25(OH)2D3 are crucial for allowing proper trophoblast invasion of the uterus without triggering a maternal immune response. Decidual natural killer cells isolated from the first-trimester decidua show decreased synthesis of cytokines such as tumor necrosis factor and interleukin-6 in response to 1,25(OH)2D3 (Evans et al. 2006). A role of 1,25(OH)2D3 as an activator of innate immunity in the placenta has been suggested. In addition to upregulation of 1α(OH)ase expression, recent findings indicate that the increased bioavailability of 1,25(OH)2D3 at the feto-maternal interface may also be partially attributed to a decrease in activity of the catabolic 24(OH)ase enzyme in the placenta. All these data suggest the role of vitamin D in the placenta to regulate both acquired and innate immune responses and a possible role for 1,25(OH)2D3 in the immunoregulation of placental development. Recent data promises a role of vitamin D for the pregnancy and the placenta. The placenta produces and is regulated by vitamin D in the process of implantation, cytokine production, and feto-maternal immune interaction (Shin et al. 2010)

DNA methylation of the placental genes (1alpha-hydroxylase, CYP27B1; vitamin D receptor, VDR; retinoid X receptor, RXR) is investigated in term placental tissue of the women who became preeclamptic ($n = 11$). Hypermethylation of genes associated with vitamin D metabolism such as CYP27B1 and VDR was detected which may influence the process of the placentation by altering the availability of the vitamin D at the feto-maternal axis (Anderson et al. 2015). Effects of active vitamin D metabolite, 1,25-dihydroxyvitamin D3 (1,25-D3), on primary human EVT isolated from first-trimester pregnancies show an induction of mRNA encoding vitamin D-responsive genes, 24-hydroxylase (CYP24A1), and cathelicidin following 1,25-D3 treatment. EVT invasion is induced with concomitant increase by MMP2 and MMP9 secretion both by 1,25-D3 and 25-D3 in vitro (Chan et al. 2015). Emerging evidences show that vitamin D influences neuronal differentiation, endocrine function, and fetus brain growth and development. Vitamin D deficiency in pregnancy alters brain development of the offspring in animals. Association of maternal vitamin (D 25(OH)) status during pregnancy ($n = 850$; >30 week)with proxies of offspring neurodevelopmental outcomes up to age of 22 years showed lack of association of vitamin D biomarker with selected brain activity parameters (Strom et al. 2014).

Brown Norway rats are best suited for vitamin D deficiency study. Analysis of the active metabolite 1,25-(OH)2-D showed a twofold increase in pregnant Sprague-Dawley (SD) and Lewis (LW) rats but a nearly tenfold decrease in pregnant BN rats compared with nonpregnant controls. BN rat had a pregnancy-dependent upregulation of CYP24a1 expression, a key enzyme that inactivates vitamin D metabolites. Finally, supplementation with 1,25-(OH)2-D significantly improved the reproductive phenotype of BN rats by increasing litter size and maternal-fetal weight outcomes (Goyal et al. 2014). Maternal vitamin D deficiency has been linked with fetal growth restriction due to increased risk of altered placental vascular pathology. Maternal pathologies were identified on placental examinations, including evidence of placental abruption, infarction, hypoxia, decidua vasculopathy, or thrombosis of fetal vessels ($n = 240$ cases). Maternal 25 (OH)D and pathology in mothers is associated with birth weight of female offspring when mother's 25(OH)D concentration is <30 nmol/L (Gernand et al. 2013). Maternal vitamin D deficiency is associated with gestational diabetes mellitus (GDM) due to differences in the placental production of vitamin D receptor (VDR). In a small pilot study with healthy ($n = 40$) and GDM ($n = 20$), approximately 27.5 % of normal pregnant women and 85 % of women with GDM had vitamin D deficiency, with serum 25(OH)D levels <20 ng/mL. Decreased serum level of 25(OH)D and elevated activity of CYP24A1 in the placenta play a key role in the development of vitamin D deficiency in GDM (Cho et al. 2013). Altered vitamin D metabolism takes place in preeclamptic placenta due to increased oxidative stress. Disruption of the activities of the vitamin D metabolic activities is evidenced due to downregulation of VDBP, CYP2R1, and VDR and upregulation of CYP27B1 and CYP24A1 protein expression in PE placenta (Ma et al. 2012). In a prospective longitudinal study ($n = 424$), lower maternal 25-hydroxyvitamin as early as 19 weeks is associated with greater femoral metaphyseal cross-sectional area. Maternal vitamin D deficiency during pregnancy as early as 19 weeks' gestation influences fetal femoral development and offspring's bone health in later life (Mahon et al. 2010).

The relationship between low vitamin D and adverse maternal outcomes such as pregnancy-induced hypertension, high blood pressure in diabetic pregnancy, gestational diabetes mellitus, recurrent pregnancy loss, and preterm delivery has been documented (Mithal and Kalra 2014). Vitamin D deficiency is common worldwide in all sections of people, but epidemiological studies reveal the high prevalence of vitamin D deficiency in women, including antenatal and lactating mothers. A vitamin D requirement is higher during second and third trimesters of pregnancy, due to its immunomodulatory functional abilities. While 1,25(OH)2D levels do not correlate directly with 25 hydroxy vitamin D levels, the physiological rise in the active metabolite, the enhanced intestinal calcium absorption, and enhanced fetal requirement of calcium (250 mg/day in the third trimester) all point to the importance of vitamin D biology in pregnancy. The deficiency of 25-hydroxyvitamin D is widely distributed among the population that included adults, pregnant women, postmenopausal women, children, and adolescents from urban and rural and affluent and non-affluent sectors with a mean circulating 25(OH)D levels between 34.8 and 46.3 nmol/L (Jani et al. 2014). The cross-sectional study reported very low level (<30.00 ng/ml) of 25(OH) D in pregnant women ($n = 150$). Widespread 25-hydroxyvitamin D deficiency in affluent and non-affluent pregnant Indian women in the land of sunshine is being highlighted recently (Jani et al. 2014). The causative factors for this deficiency seem to be multifactorial that include sun exposure index, family income, skin type, dietary phytate and calcium, and low intake of vitamin D from the diet. Vitamin D deficiency is associated with antiphospholipid syndrome (APS) and preeclampsia. Pregnant women with APS are at increased risk of recurrent pregnancy loss (RPL) and preeclampsia. Antiphospholipid antibodies (aPL) directly regulates trophoblast function. Vitamin D attenuates aPL-induced trophoblast inflammatory response in the HTR8 cells and primary cultures (Gysler et al. 2015). Epidemiological studies established a positive association between low maternal vitamin D status in pregnancy and the incidence of preeclampsia and suggest that the deficiency may be an independent risk factor for preeclampsia (Bodnar et al. 2007; Wei et al. 2012). Whether maternal plasma level of 25-hydroxyvitamin D [25(OH)D] is associated with angiogenesis and endothelial dysfunction is investigated at 12–18 and 24–26 weeks of gestation in pregnant women ($n = 697$) and sFlt-1, PIGF, ICAM-1, and VCAM-1 levels at 24–26 weeks, respectively (Wei et al. 2013). Maternal PIGF levels are significantly lower in women with 25(OH)D less than 50 nmol/L at both time points. Both maternal 25(OH)D and PlGF levels are inversely associated with the risk of preeclampsia, but is not associated with impaired angiogenesis pathway. Further studies with placental angiogenesis are required since preeclampsia is a disease due to insufficient angiogenesis of the placenta. Lower vitamin D levels at the first trimester are associated with higher risk of developing gestational diabetes mellitus (Lacroix et al. 2014). It suggests that first-trimester 25-hydroxyvitamin D (25OHD) levels are an independent risk factor for developing GDM and are associated with insulin resistance at the second trimester. The relationship of lower vitamin D level and IL-6 is compared in preeclampsia ($n = 100$) and normotensive pregnant women ($n = 100$). The mean plasma level of 25-hydroxyvitamin D is 49.4 ± 22.6 nmol/L in

Table 6.2 Maternal vitamin D deficiency on the placental anomalies

Effect of vitamin D deficiency	References
Suboptimal levels of vitamin D and omega-3 fatty acids in early development lead to dysfunction in serotonin activation and alter brain function	Patrick and Ames (2015)
Down regulation of CYP2R1, VDR, and upregulation of CYP27B1, CYP24A1 protein expression in PE placenta	Ma et al. (2012)
Elevated activity of CYP24A1 in the placenta with serum 25(OH)D levels <20 ng/mL links with GDM	Cho et al. (2013)
Hypermethylation of CYP27B1, VDR genes in term placenta of PE	Anderson et al. (2015)
Maternal 25(OH)D serum level <30 nmol/L is associated with LBW of female offspring	Gernand et al. (2013)
Maternal vitamin D deficiency as early as 19 weeks' gestation alters fetal femoral development and offspring's bone health in later life	Mahon et al. (2010)
Maternal PIGF levels significantly lower in pregnant women with <50 nmol/L 25(OH)D	Wei et al. (2013)

normotensives and 42.3 ± 17.3 nmol/L in preeclamptic women ($P = .01$). A significant association between vitamin D deficiency and preeclampsia but not with IL-6 suggests vitamin D deficiency alters the pathogenesis of preeclampsia independent of IL-6-mediated inflammatory pathways (Xu et al. 2014). Dietary vitamin D restriction in pregnant mice is associated with maternal hypertension and altered placental and fetal development (Liu et al. 2013).

All these data suggest the requirement of vitamin D to pregnant women who are deficient in vitamin D during pregnancy. Deficiency of vitamin D during pregnancy has several effects on placental function (Table 6.2). Data proves the safety and efficacy of 4000 IU vitamin D, when administered daily over 6 months of pregnancy (Hollis et al. 2011). However, supplementation is not associated with maternal vitamin D level and birth weight. Simultaneously, no adverse event due to vitamin D supplementation is documented. The duration of the study (from 12 weeks of gestation onward), the dose used (400, 2000, and 4000 IU daily), a control group supplemented with 400 IU/day, and the large subject size are the highlight of this clinical trial. Similar data is obtained from Arab women where vitamin D of 2000 and 4000 IU/day, from 12 to 16 weeks, is supplemented to antenatal women from vitamin D-deficient regions (Dawodu et al. 2013). Study from New Zealand has proven the safety and utility of supplementing 2000 IU/day of vitamin D from 27 weeks onward and continuing 800 IU/day supplementation in infants till 6 months of age (Grant et al. 2014). Supplement of vitamin D with omega-3 fatty acids improve cognitive function and behavior of brain disorders. It is known that serotonin synthesis, release, and function in the brain are modulated by vitamin D and omega-3 fatty acids specially eicosapentaenoic acid (EPA) and docosahexaenoic acid (DHA). Inadequate levels of vitamin D and omega-3 fatty acids are common (70 % of general population) including pregnancy that indicates suboptimal synthesis of brain serotonin. Brain serotonin is synthesized from tryptophan by tryptophan hydroxylase 2 which is transcriptionally activated by vitamin

D hormone. It is possible that suboptimal levels of vitamin D and omega-3 fatty acids during early development lead to dysfunction in serotonin activation and thereby alters function of brain development (Patrick and Ames 2015).

6.3 Vitamin A and Developmental Pathways

Vitamin A (retinol) belongs to the "retinoid" family that includes biological active compounds retinol (ROH) and the many synthetic analogues to the retinol, with or without a biological activity. Pro-vitamin A is the dietary source of retinol and is supplied as carotenoids (mainly β-carotene) in vegetables and preformed retinyl esters (long-chain fatty acid esters of retinol such as palmitate, oleate, stearate, and linoleate) in animal meat (Blomhoff 1994). Two common pathways exist for mammalian blood transport of retinoids. First, the retinyl esters and carotenoids can be incorporated in intact or remnant chylomicrons or very low-density lipoproteins (Debier and Larondelle 2005). Second is the binding of retinoid with a specific binding protein known as retinol binding protein (RBP) which is pre-complexed with transthyretin. The formation of ternary complex is crucial in retaining retinoid since the complex prevents glomerular filtration of small RBP retinol form and increase the affinity of RBP for retinol. Human fetus is not capable to synthesize retinoid de novo and therefore totally dependent on maternal vitamin A supply. Hepatic vitamin A is detected after birth further confirms that placental transport remains active to transfer this vitamin from mother to newborn. Depending on the isomers, rate of transfer of retinoids are varied across the placenta (Ross and Gardner 1994). It has been proposed that variability in teratogenic effects of comparable amounts of vit A could be due to different maternally absorbed retinoids. Vitamin A is provided to the fetus through a limited and tightly controlled placental transfer. The amount of retinol provided to the fetus is usually maintained constantly until maternal stores are depleted. The first step of retinol transfer from maternal blood to the embryo involves the binding of RBP to retinol. Since RBP of fetal origin is unable to cross the placenta and cannot get entry into the maternal circulation, therefore, it dissociates from maternal RBP after the binding to its receptor at the maternal face of placental barrier. Bound to the cRBPs, the retinol passes through the cytoplasm of the trophoblastic cells and enters the fetal circulation, where a new complex is formed using transthyretin and RBP of fetal origin (Quadro et al. 2005). Retinol bound to RBP is not the only source for retinoids reaching the embryo. Results demonstrate that retinyl esters in lipoprotein particles can be a significant source for retinoids (present postprandial in maternal blood) and can be used by the fetus to support embryogenesis. These data established that very low-density lipoproteins and low-density lipoproteins can be taken up by the placental cells. This postprandial pathway may be also the delivery pathway of carotenoids (precursors of vitamin A) to the fetus through the placenta. It is known that the placenta acts as a transitory, primitive functional liver during the first stages of mammalian development. During this period, the placenta stores retinol, till the time the liver matures and is efficient to take up and secrete RBP. During early organogenesis, the retinyl ester content of the placenta is nearly eightfold higher than the embryonic content. At the end of gestation, the embryonic retinyl ester content is nearly fourfold greater than placental one. Abnormalities concerning this switch between placental and embryonic retinyl esters stores are associated with intrauterine growth retardations (Marceau et al. 2007).

Vitamin A deficiency and excess have profound effects on the development of the vertebrate embryo. A molecular basis for these phenomena suggests that vitamin A in its active forms, RAs, act through ligand-activated transcription factors and that RA induces a change in the expression pattern of homeobox genes clusters. Prenatal malnutrition, particularly vitamin A deficiency during critical periods of fetal development, represents a plausible cause sensorineural hearing loss (Emmett and West 2014). Vitamin A, in the form of its active metabolite retinoic acid (RA), has been shown to be indispensible for inner ear development. Retinoic acid acts as a differentiation factor in the development of hindbrain, and the downstream factors regulated by RA demonstrate the essentiality of vitamin A in inner ear formation. The amount of supplemental vitamin A is needed to be consumed in a deficient population to protect ear development is not known (Institute of Medicine Panel on Micronutrients 2001). High dietary intake of preformed vitamin A appears to be teratogenic (Rothman et al. 1995). Retinoic acids (13-cis and 13-trans) are known teratogens, and their precursor is retinol, a form of vitamin A. In 1995, Rothman et al. demonstrated an association between excessive vitamin A, >10,000 IU/day, during the first trimester of pregnancy and teratogenic effects, particularly in the central nervous system. However, vitamin A deficiency has long been known to be deleterious to the mother and fetus. The excess 13-trans retinoic acid derived from metabolized beta-carotene in the fetus increases the concentration of the more teratogenic 13-cis retinoic acid due to the shift in equilibrium of isomer formation. Therefore, it is important to monitor all potential sources of fetal 13-cis and 13-trans retinoic acid, including nutritional supplements, dietary retinol, and beta-carotene, particularly in the first trimester of pregnancy (Goldberg 2011). Vitamin A supplement can improve wound healing by decreasing fibrosis and increasing transforming growth factor-beta (TFG-beta) in pregnancy-related infections in vitro. During infection, vitamin A acts as an immune enhancer by increasing the adequacy of natural killer (NK) cells and increasing antibody production. Pro-vitamin A in the form of beta-carotene can act as an antioxidant by decreasing endothelial cell damage and promoting the vasodilator effect of nitric oxide that possibly improve better outcome of pre-eclampsia in pregnancy. It should be noted that all these effects mediated by vitamin A directly or indirectly can only be restricted for obstetric hemorrhage, anemia in pregnancy, hypertension in pregnancy, and pregnancy-related infections (Faisel and Pittrof 2000).

The threshold harmful level of vitamin A to act as teratogen in pregnancy has not been established by epidemiological studies. Considering its possible adverse effect in higher doses during pregnancy, evidences suggest that vitamin A intake below 3000 μg (10,000 IU)/day is safe; however, later studies suggest that vitamin A intake up to 9000 μg (30,000 IU)/day represents safe levels as well (Miller et al. 1998; Hartmann et al. 2005). A population-based case-control study from Norway evaluates the association between maternal intake of vitamin A from diet and supplements and risk of having a baby with an orofacial cleft from the data on maternal dietary intake of cases ($n = 535$) and controls ($n = 693$). Maternal intake of vitamin A is not only associated with reduced risk of cleft palate, but the study could not find any evidence of increased risk of clefts among women in the study with the highest 5 % of vitamin A intake (Johansen et al. 2008) . Thus, a protective association between high maternal intake of vitamin A and risk of cleft palate of the offspring seems to suggest a new role for vitamin A in the pathogenesis of the cleft palate. The amount of vitamin A intake through supplementation and diet is always measured through the lenses of its teratogenicity. Compared to the study reported by Rothman et al. (1995), in which increased maternal retinol intake (4500 μg: 1500 μg = defects vs. normal) from diet and supplements is associated with the of risk cranial neural-crest tissue, the mean retinol intake from supplements (477 μg) accounts for approximately one-third of the total intake of retinol (1210 μg) (Johansen et al. 2008).

Maternal vitamin A supplement causes a reduction in the night blindness, mortality related to pregnancy, and mortality of infants born to mothers prone to night blindness. A randomized cohort of children born to women randomly assigned to receive either supplemental vitamin A (7000 μg retinol equivalents) or placebo weekly during a continuous 3.5-year study from preconception to postpartum did not improve cognition or motor development at ages 10–13 years in an undernourished setting in Nepal (Buckley et al. 2013). A cross-sectional study investigated maternal and neonatal factors with serum vitamin A concentration in matched mother-newborn pairs ($n = 100$) at the semi-urban setup of Northern India (Agarwal et al. 2008). Low neonatal vitamin A levels is associated with prematurity and intrauterine growth retardation, and the effect is influenced by other associated factors such as lack of antenatal care, lower maternal hemoglobin levels, and reduced placental weight. More studies are required on vitamin A supplementation in pregnant women and preterm neonates possibly in the similar nutritional background. A recent study postulates that the placenta could be used as a possible predictor of vitamin A deficiency (Gomes et al. 2010). The positive association is established between retinol concentration from the mother and umbilical cord and placental concentration of retinol and carotenoids using puerperal women ($n = 262$) and their newborns.

6.4 Vitamin E and Placentation

Vitamin E is originally discovered as an essential factor for reproduction. Isolation of vitamin E by Evans and Bishop is evidenced first time with their wheat germ feeding experiment in rats. Vitamin E is named as tocopherol, from the Greek words "tocos" (birth), and "pherin" (to carry) with the ending "-ol" indicating its status as a chemical alcohol. Vitamin E comprises eight isomers α-, β -, γ-, and δ- tocopherols and the α-, β -, γ-, and δ- tocotrienol occur naturally as RRR-stereoisomers. RRR-α-tocopherol is selectively enriched by the liver in plasma by the α-tocopherol transfer protein (α-TTP), and remaining isoforms are not enriched in plasma (therefore level below 1 μM) but metabolized. Mean plasma value of α-tocopherol is normally about 23.2 μM with a plasma level below 11.6 μM regarded as deficient. In human, premature babies are often characterized by vitamin E deficiency, with plasma α-tocopherol levels lower than 5 μg/mL (11.6 μM).

The prominent understanding of vitamin E action as antioxidant is well established. Generally, during vitamin E deficiency, lipid peroxidation occurs which leads to oxidative chain reactions with a production of free radicals that ultimately participate in nonspecific reactions with functional and structural compounds. Vitamin E can reduce lipid peroxides or scavenge free radicals from chain reactions. The role of vitamin E is similar with the enzyme glutathione peroxidase system for their antioxidative action. However, antioxidative action of vitamin E and glutathione peroxidase operates in different subcellular location, e.g., glutathione peroxidase operates in cytosol, whereas vitamin E is mainly located in the membranes of the cell (Veen and Grimbergen 1975). Interestingly, fat-soluble vitamin E is the least accumulated vitamin among the fat-soluble vitamins A, D, and K. The need for vitamin E is increased during pregnancy when the diet is rich in PUFA. Increased amounts of vitamin E are needed in low-birth-weight (LBW) infants fed with diets rich in PUFA. Accumulation of lipid-soluble vitamins in fetal tissue occurs mostly in the third trimester of gestation. Due to this, preterm infant stores less of vitamin E as compared to full term. It could also possible that lipid-soluble vitamins are mostly utilized during early trimesters of pregnancy to provide optimum nourishment for the feto-placental unit (Orzalesi and Colarizi 1982).

α-TTP is a cytosolic protein mostly present in the liver that specifically binds α-tocopherol (Dutta-Roy 1997). Expression of α-TTP is increased after implantation. Data shows that uterine α-TTP may contribute an important role in supplying vitamin E. Mice placentas of pregnant α-TTP(-/-) show disrupted growth of labyrinthine trophoblasts which is possibly due to inadequate supply of antioxidant vitamin E. Excess α-tocopherol prevented placental failure and allowed full-term pregnancies (Jishage et al. 2001). α-TTP is expressed in human placenta in different subcellular compartment that includes in cytosol but predominantly in the nuclear fraction of the trophoblast and endothelium fetal capillaries. α-TTP knockout mice are infertile and exhibit similar symptom as with severe vitamin E deficiency (Muller-Schmehl et al. 2004). Vitamin E is critically required during the early stages of life that span from conception to the postnatal development of the offspring. The mechanism of the uptake of this vitamin is probably associated with lipoprotein metabolism of the mother and placenta. In addition to this, α-TTP plays a specific role to transfer this vitamin across the placenta. Lower level of α-TTP is found in the fetus as compared to maternal circulation. Colostrum contains very high levels of vitamin E. Therefore vitamin E acts as an essential defense against oxygen toxicity of the newborn. Since this vitamin E accumulates mostly in third trimester, preterm infants are susceptible to oxidative stress and require longer time to attain adequate level of α-tocopherol than full-term infant. In contrary, vitamin E as therapeutic agent for PE did not find any beneficial effect that indicates possible involvement of other associative factors with vitamin E (Debier 2007).

Effect of vitamin E on angiogenesis and vasculogenesis of the placenta of pregnant ewes suggest possible involvement of VEGF expression. Impaired vasculogenesis and angiogenesis of the placenta due to decreased expression of VEGF lead to placental ischemia, elevated free radicals, and impaired fetal nutrient supply that make the fetus vitamin E deficient. An increase in placental angiogenesis and vascular network formation is detected in pregnant ewes supplemented with vitamin E due to increased VEGF expression (Kasimanickam et al. 2010). Recent evidence suggests that activation of alpha-tocopherol could attribute to active form of vitamin E that induces angiogenesis and vasculogenesis in the placenta. Several reports are available on regulation of altered VEGF expression by vitamin E, e.g., phosphorylated form of aTP induces VEGF expression at the mRNA and protein level leading to increase in vitro angiogenesis in HUVEC (Zingg et al. 2012). The phosphorylated form of α-tocopherol, α-tocopheryl phosphate, increases the expression of VEGF. The stimulatory effect of α-tocopherol on angiogenesis and vasculogenesis is potentiated by phosphorylation to α-tocopherol, which may act as a cofactor or active lipid mediator increasing VEGF expression. One possible hypothesis on the involvement of vitamin E on angiogenesis of the

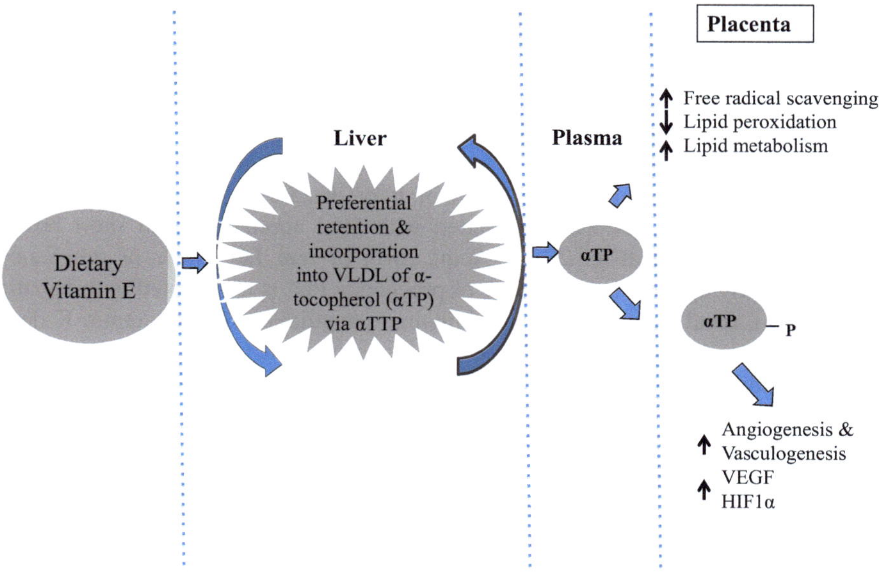

Fig. 6.1 Vitamin E absorption and its roles in angiogenesis during placentation. α-tocopherol (αTP), α-tocopherol transport protein (αTTP). Adapted from Dutta-Roy et al. 1994

placenta is that induction of VEGF by α-tocopherol could be mediated as a result of the stimulation of the PI3k/Akt signaling pathway that may lead to activate angiogenesis and vasculogenesis of the placenta (Fig. 6.1). It is important to understand that trophoblast development and placentation occurs differently across the species, and therefore the expression of tocopherol transport genes or vitamin E deficiency effects in animal may not be similar with human.

6.5 Maternal Vitamin K Deficiency and Altered Fetus Development

Deficiency of vitamin K in pregnant women could pose a risk for the newborn. Vitamin K deficiency is critical for the newborn since they become deficient after birth. Vitamin K is essential for the activation of coagulation factors VII, IX, X, and prothrombin. Both human milk and newborn contains less stores of vitamin K. Vitamin K deficiency and hemorrhage is a wide problem across the globe. The deficiency exposes the fetus to several rare bleeding disorders including brain damage from intracranial hemorrhage. The deficiency can be restored with a single intramuscular injection of vitamin K immediately after birth. In some instances, a fear of risk with this dose is reported. However, subsequent studies nullify this assumption (McCarthy 2014; Gosai et al. 2014; Arya et al. 2015).

A flat facial profile popularly known as "binder phenotype" is consistently reported with vitamin K deficiency from hyperemesis gravidarum (HG) (Lane et al. 2015). HG is a severe form of nausea and vomiting during pregnancy. Altered liver function and decreased prothrombin time are common feature that is secondarily linked with vitamin K deficiency of the pregnant women. Evaluating early vitamin K status before pregnancy may be useful in preventing vitamin K deficiency-associated embryopathy from HG (Chraibi et al. 2015). Deficiency of vitamin K is widely reported in cases of liver disease. Maternal deficiency of fat-soluble vitamins has been associated with fetal anomalies with short bowel syndrome (Grabosch et al. 2013). Acute form of rare fatty liver of pregnancy (AFLP) mostly appears in later stage of pregnancy that is manifested with hepatic dysfunction, renal insufficiency, and impaired coagulation. Early vitamin K deficiency and bleeding in fetus are linked with maternal acute fatty liver of pregnancy. Early vitamin K deficiency bleeding in a neonate is associated with maternal Crohn's disease (Ohishi et al. 2014). Neonatal intracranial bleedings and birth defects are reported in pregnant women with previous bariatric surgery. Maternal deficiency of vitamin K_1 is often reported under this condition (Jans et al. 2014). During the first trimester, pregnant women have low K_1 serum levels (<0.8 nmol/l), and the level further goes down in women with previous bariatric surgery (.44 versus .64 nmol/l; $P = .016$).

Supplementation with vitamin K_1 during pregnancy is effective in restoring vitamin K_1 level and can protect fetus against the intracranial hemorrhage.

Accurate marker for vitamin K deficiency in term fetus is not available to find subclinical deficiency. Proteins induced by vitamin K absence (PIVKA-II) are the inactive forms of clotting factors. Plasma levels of PIVKA-II in term infants have been shown as a sensitive marker of subclinical vitamin K deficiency over prothrombin time (Dituri et al. 2012).

6.6 Vitamin C and Placentation

Vitamin C is known for its antioxidant effects by maintaining redox homeostasis. Developing brain is highly susceptible to oxidative damage due to absence of mature brain, decreased level of antioxidant enzymes, high oxygen consumption, and oxidation-prone PUFAs. Vitamin C (ascorbic acid) acts as cofactor in the formation and reuptake of the neurotransmitter of the brains. Low maternal intake of vitamin C is associated with preeclampsia, preterm, PROM, and decreased birth weight. The impact of vitamin C deficiency in utero fetal development especially on brain development is reported recently. Vitamin C deficiency during pregnancy leads to hippocampal volume reduction and impaired neuronal migration of the animal model. Animal study shows that distribution of plasma vitamin C between mother and the fetus become altered during maternal vitamin C deficiency. Compared with maternal plasma vitamin C, plasma vitamin C of newborn pups remained lower in the vitamin C-deficient group. Preferential transport of vitamin C from the mother to the fetus is superseded by the placenta during sustained maternal vitamin C deficiency. It is possible that the placenta retains vitamin C in order to ensure adequate fetal level in case of severe maternal deficiency of vitamin C (Schjoldager et al. 2013). Placental uptake of vitamin C requires functional SVCT2 transporters. Human trophoblast cell line HTR-8/SVneo acts as a suitable model to investigate vitamin C transport in the trophoblast (Biondi et al. 2007). Placental sodium-dependent vitamin C transporters (SVCT) 1 and 2 are expressed in the first trimester with predominant expression of SVCT2 mRNA.

Maternal vitamin C stores that reach below a critical level of 18 μg/10(8) leukocytes has been suggested to be linked with altered collagen synthesis that may lead to premature rapture of membrane (PROM). In spite of multiple etiologies, preterm birth is usually preceded by PPROM. PPROM contributes 40 % of preterm birth and represents a broad consensus pathway to preterm birth. Vitamin C deficiency contributes one of the key factors of PROM; studies observed that the rupture of chorioamniotic membranes is linked with altered turnover of collagen synthesis that is induced by an imbalance between synthesis and degradation of various MMPs (Casanueva et al. 2005); vitamin C (ascorbic acid) acts as a cofactor that regulates the posttranslational regulation of MMPs. MMP9 and other MMP activities are increased in preterm birth (Sundrani et al. 2013). Prolactin receptor (PRL-R) expression is increased in chorioamniotic membranes, decidua, and placenta at the time of delivery. PRL regulates synthesis of some MMPs (MMP-2, MMP-9) by modulating their expression. Study shows that vitamin C increases the inhibition of

PRL in animal model. Vitamin C deficiency induces the expression/activity of extracellular MMP. All these data suggest vitamin C deficiency could be a risk factor for preterm birth (vitamin C excreted potent effects in vitro when first-trimester placenta is exposed to xenobiotics-carcinogens such as mercury or other toxic metal). Placental protective enzyme quinone reductase (QR) is induced by vitamin C which indicates that vitamin C (ascorbic acid) plays a protective role for the fetus when the mother is exposed to adverse environment (Barnea et al. 1993). Placental protein 13 (PP13) is a placenta-specific protein biomarker in the first trimester which is decreased in the maternal serum level of the PE. Vitamin C, dose dependently, increases this protein in vitro which suggests a point that early vitamin C treatment in the early first trimester of pregnancy may be useful in case of PE (Orendi et al. 2010).

6.7 Impact of Mineral Nutrition and Possible Role in Placentation

Since conception to birth, an offspring is completely dependent on the nutrition of the mother via placenta. The priority problems of the intervention are a must to include supplementation with iron-folic acid or multiple micronutrients (Mason et al. 2014).

Iron deficiency (ID) is the most common form of nutrient deficiency, affecting two billion people and 30 % of pregnant women and their offspring. Iron deficiency is associated with poor pregnancy outcome and preterm delivery. During pregnancy, there is a significant demand in iron requirements due to increased red cell mass and feto-placental growth. Iron deficiency and anemia during pregnancy is a global burden and is closely related with maternal and perinatal mortality and brain development index. A study of pregnant women ($n = 201$) from Turkey where the Mediterranean Cuisine is followed with fresh fruit and vegetables concluded a wide prevalence of iron deficiencies in the pregnant women. Based on the WHO guideline of anemia (hemoglobin, <11 g/dl, and iron deficiency as ferritin, <15 μg/L), in the present study, 40.3 % of women had depleted iron stores and 4.5 % had iron deficiency anemia (Karabulut et al. 2011). Nutrition-related iron deficiency is the main cause of anemia throughout the world. The first-trimester pregnancies without any supplementation more likely reflect the preconception status. This result suggests iron deficiency is a common problem in these vegetarian populations. The high prevalence of iron deficiencies is detected in the first trimester of pregnancy that seriously poses reproductive risks for feto-placental development. The learning and memory deficit occur where iron is deficient and remained persistent even after repletion. The developing hippocampus is especially susceptible to early ID (Fretham et al. 2011). Iron is essential for hemoglobin synthesis. Iron deficiency is widespread in 50 % of all pregnant women in developing countries. Deficiency induces maternal and fetal stress, by increasing cortisol and corticosteroid hormone

(CRH) that leads to IUGR and possible fetal growth. The number of studies has shown that maternal iron deficiency in pregnancy has an adverse effect on birth outcome and many have shown that maternal iron supplementation increases birth size and minimizes risk of prematurity and the risk of other problems at birth. Data suggests decreased iron status is associated with increased risk of developmental delay and increased incidence of mental problems in adulthood (McArdle et al. 2011). Deficiency of iron (ID) in combination with $n-3$ PUFA ($n-3$ FAD) disrupts brain monoamine level and cognitive performance of the double deficient rat and induces more memory deficit as compared to ID or $n-3$ FAD alone. Combined deficiency of iron and n-PUFA is not uncommon in low-income and vegetarian countries where children born to a mother with poor $n-3$ PUFA status in their diet with low iron and low omega-3 fatty acid. Iron is a cofactor of the enzymes responsible for the conversion of alpha-linolenic acid (ALA) to EPA and DHA. The possibility of combined intake of diet rich in ALA and Fe might be effective to restore such inadequacy during critical phase of the development including early pregnancy (Baumgartner et al. 2014). A prospective cohort study in the UK investigates the association between iron intake in pregnancy and size at birth of pregnant of women ($n=1274$) at 12-week gestation by dietary recall method. A total 80 % reported dietary iron intake below the UK reference nutrient intake of 14.8 mg/day. Those reported taking iron-containing supplements in the first, second, and third trimesters are 24, 15, and 8 %, respectively. A positive relationship exists between total iron intake, from food and supplements, in early pregnancy and birth weight. Iron intake, both from diet and supplements, during the first trimester of pregnancy is higher in vegetarians and women with a better socioeconomic profile. In India, maternal anemia and iron deficiency anemia widely exists with pregnant women. Recent data of pregnant women ($n=366$) reported that anemia and iron deficiency are common in pregnancy. Data shows about 30 % of women are anemic (Hb < 11.0 g/dL), 40 % are iron deficient (ferritin < 12 μg/L), and 22 % are categorized as the case of iron deficiency anemia (Finkelstein et al. 2014).

Calcium plays an important role in muscle function, blood vessel dynamics, and nerve impulse transmission. Calcium is required for the normal blood coagulation. Deficiency in pregnancy is not common in pregnancy. Calcium [Ca (2+)] actively crosses the trophoblast layer up to 30 g during human pregnancy. Calcium actively plays a role for skeletal development of the growing fetus. Most importantly, it acts as an essential intracellular second messenger in regulating many cellular processes such as gene transcription, proliferation, differentiation, and signal transduction pathway in placental function. Ca2+ wave has been shown to induce fertilization of the oocyte. The intracellular Ca2+ concentration is key for the blastocyst implantation and proper placental development and function (Baczyk et al. 2011).

Magnesium acts as an enzyme cofactor and activators in several biochemical pathways. Hypomagnesia in diabetic women leads to maternal hypomagnesia and fetus hypocalcaemia that may interfere in fetal growth and development. In addition to its antioxidant function, zinc (Zn) supports the immune system. Zinc is essential for embryogenesis and important for normal fetal growth. The deficiency of Zn in the first trimester could be the risk factor for PE. Selenium is the antioxidant that supports humoral and cell-mediated immunity and contributes an important role in reproduction. First-trimester dietary intake is not associated with gestational hypertension in a recent prospective study (Tande et al. 2013). However, women with low serum zinc levels in early pregnancy are at increased risk for the development of gestational hypertension indicating that maternal nutritional status in early pregnancy represents a critical period for placental and fetal development, serving as a target for interventions to improve pregnancy outcomes. Increased level of plasma copper and iron and decreased level of manganese, selenium, and zinc are associated with preeclampsia (PE). Increased oxidative stress and altered trace elements in maternal plasma is the indicative of disturbance in the antioxidative status during pregnancy. In a longitudinal study Type-1 diabetic pregnant women who later become preeclamptic, only plasma zinc was significantly higher at the first trimester, while copper/zinc and copper/high-density lipoprotein cholesterol ratios were higher throughout gestation. Higher ratio of copper to zinc in the first trimester is thought to contribute oxidative stress in those with type 1 diabetes who later become preeclamptic (Basu et al. 2015). Antioxidant status is altered in women who have had a miscarriage. Serum analysis is observed with lower antioxidative status in pregnant women ($n = 83$) with miscarriages as compared to control group ($n = 35$) in their first trimester of pregnancy. Total antioxidant and Cu level are lower, whereas Mn in serum are higher in miscarriages under this condition (Omeljaniuk et al. 2015) Antioxidant supplements to the pregnancy disorders yet to find a successful measure to combat diseases where reproductive performance, including infertility, miscarriage, diabetes-related congenital malformations, and preeclampsia where balance between oxidative stress and antioxidant are failed. Early supplementation with vitamins C and E could not prevent PE in women at risk of preeclampsia. Potent combination of antioxidant vitamins, much earlier intervention (preconception), and inclusion of selenium in combination with statin are advocated (Poston et al. 2011). Recent data suggest significant benefit of multiple micronutrient supplementation during pregnancy in reducing low birth weight, small for gestational age births as compared to the usual iron-folate supplements (Zerfu and Ayele 2013).

6.8 Summary

Preconception requirements of fat- and water-soluble vitamins are important for placentation processes. No safety data, however, is available for the first trimester with this dose of these vitamins. Several in vitro and epidemiological studies suggest their importance in feto-placental unit. However, more clinical data is needed to measure the effect of these vitamins and micronutrient feto-placental growth and their effects on birth weight, neonatal health and neurodevelopment, allergy response, and maternal outcomes.

References

Agarwal, K., Dabke, A. T., Phuljhele, N. L., & Khandwal, O. P. (2008). Factors affecting serum VITAMIN A levels in matched maternal-cord pairs. *Indian Journal of Pediatrics, 75*, 443–446.

Anderson, C. M., Ralph, J. L., Johnson, L., Scheett, A., Wright, M. L., Taylor, J. Y., Ohm, J. E., & Uthus, E. (2015). First trimester vitamin D status and placental epigenomics in preeclampsia among Northern Plains primiparas. *Life Sciences, 129*, 10–15.

Arya, S., Richardson, C. J., Jain, S., & Swischuck, L. E. (2015). Early vitamin K deficiency bleeding in a neonate associated with maternal acute fatty liver of pregnancy. *AJP Reports, 5*, e193–e195.

Baczyk, D., Kingdom, J. C., & Uhlen, P. (2011). Calcium signaling in placenta. *Cell Calcium, 49*, 350–356.

Barnea, E. R., Avigdor, S., Boadi, W. Y., & Check, J. H. (1993). Effect of xenobiotics on quinone reductase activity in first trimester explants. *Human Reproduction, 8*, 102–106.

Basu, A., Yu, J. Y., Jenkins, A. J., Nankervis, A. J., Hanssen, K. F., Henriksen, T., Lorentzen, B., Garg, S. K., Menard, M. K., Hammad, S. M., Scardo, J. A., Aston, C. E., & Lyons, T. J. (2015). Trace elements as predictors of preeclampsia in type 1 diabetic pregnancy. *Nutrition Research, 35*, 421–430.

Baumgartner, J., Smuts, C. M., & Zimmermann, M. B. (2014). Providing male rats deficient in iron and n-3 fatty acids with iron and alpha-linolenic acid alone affects brain serotonin and cognition differently from combined provision. *Lipids in Health and Disease, 13*, 97.

Biondi, C., Pavan, B., Dalpiaz, A., Medici, S., Lunghi, L., & Vesce, F. (2007). Expression and characterization of vitamin C transporter in the human trophoblast cell line HTR-8/SVneo: Effect of steroids, flavonoids and NSAIDs. *Molecular Human Reproduction, 13*, 77–83.

Blomhoff, R. (1994). Transport and metabolism of vitamin A. *Nutrition Reviews, 52*, S13–S23.

Bodnar, L. M., Catov, J. M., Simhan, H. N., Holick, M. F., Powers, R. W., & Roberts, J. M. (2007). Maternal vitamin D deficiency increases the risk of preeclampsia. *Journal of Clinical Endocrinology and Metabolism, 92*, 3517–3522.

Buckley, G. J., Murray-Kolb, L. E., Khatry, S. K., Leclerq, S. C., Wu, L., West, K. P., Jr. & Christian, P. (2013). Cognitive and motor skills in school-aged children following maternal vitamin A supplementation during pregnancy in rural Nepal: A follow-up of a placebo-controlled, randomised cohort. *BMJ Open, 3*.

Casanueva, E., Ripoll, C., Meza-Camacho, C., Coutino, B., Ramirez-Peredo, J., & Parra, A. (2005). Possible interplay between vitamin C deficiency and prolactin in pregnant women with premature rupture of membranes: Facts and hypothesis. *Medical Hypotheses, 64*, 241–247.

Chan, S. Y., Susarla, R., Canovas, D., Vasilopoulou, E., Ohizua, O., Mccabe, C. J., Hewison, M., & Kilby, M. D. (2015). Vitamin D promotes human extravillous trophoblast invasion in vitro. *Placenta, 36*, 403–409.

Cho, G. J., Hong, S. C., Oh, M. J., & Kim, H. J. (2013). Vitamin D deficiency in gestational diabetes mellitus and the role of the placenta. *American Journal of Obstetrics and Gynecology, 209*(560), e1–e8.

Chraibi, Z., Ouldamer, L., Body, G., & Bacq, Y. (2015). Hyperemesis gravidarum: A ten-year French retrospective study of 109 patients. *Presse Médicale, 44*, e13–e22.

Christakos, S., Hewison, M., Gardner, D. G., Wagner, C. L., Sergeev, I. N., Rutten, E., Pittas, A. G., Boland, R., Ferrucci, L., & Bikle, D. D. (2013). Vitamin D: Beyond bone. *Annals of the New York Academy of Sciences, 1287*, 45–58.

Dawodu, A., Saadi, H. F., Bekdache, G., Javed, Y., Altaye, M., & Hollis, B. W. (2013). Randomized controlled trial (RCT) of vitamin D supplementation in pregnancy in a population with endemic vitamin D deficiency. *Journal of Clinical Endocrinology and Metabolism, 98*, 2337–2346.

Debier, C. (2007). Vitamin E during pre- and postnatal periods. *Vitamins and Hormones, 76*, 357–373.

Debier, C., & Larondelle, Y. (2005). Vitamins A and E: Metabolism, roles and transfer to offspring. *British Journal of Nutrition, 93*, 153–174.

Dituri, F., Buonocore, G., Pietravalle, A., Naddeo, F., Cortesi, M., Pasqualetti, P., Tataranno, M. L., & Agostino, R. (2012). PIVKA-II plasma levels as markers of subclinical vitamin K deficiency in term infants. *Journal of Maternal-Fetal and Neonatal Medicine, 25*, 1660–1663.

Dutta-Roy, A. K. (1997). alpha-tocopherol-binding proteins: Purification and characterization. *Methods in Enzymology, 282*, 278–297.

Dutta-Roy, A. K., Gordon, M. J., Campbell, F. M., Duthie, G. G., & James, W. P. T. (1994). Vitamin E requirements, transport, and metabolism: Role of α-tocopherol-binding proteins. *The Journal of Nutritional Biochemistry, 5*, 562–570.

Emmett, S. D., & West, K. P., Jr. (2014). Gestational vitamin A deficiency: A novel cause of sensorineural hearing loss in the developing world? *Medical Hypotheses, 82*, 6–10.

Evans, K. N., Nguyen, L., Chan, J., Innes, B. A., Bulmer, J. N., Kilby, M. D., & Hewison, M. (2006). Effects of 25-hydroxyvitamin D3 and 1,25-dihydroxyvitamin D3 on cytokine production by human decidual cells. *Biology of Reproduction, 75*, 816–822.

Faisel, H., & Pittrof, R. (2000). Vitamin A and causes of maternal mortality: Association and biological plausibility. *Public Health Nutrition, 3*, 321–327.

Finkelstein, J., Duggan, C., Thomas, T., Bose, B., Samuel, T., Srinivasan, K. & Kurpad, A. (2014). Maternal anemia, iron deficiency, and pregnancy outcomes in India (804.10). *The FASEB Journal, 28*.

Fretham, S. J., Carlson, E. S., & Georgieff, M. K. (2011). The role of iron in learning and memory. *Advances in Nutrition, 2*, 112–121.

Gernand, A. D., Bodnar, L. M., Klebanoff, M. A., Parks, W. T., & Simhan, H. N. (2013). Maternal serum 25-hydroxyvitamin D and placental vascular pathology in a multicenter US cohort. *American Journal of Clinical Nutrition, 98*, 383–388.

Goldberg, J. S. (2011). Monitoring maternal Beta carotene and retinol consumption may decrease the incidence of neurodevelopmental disorders in offspring. *Clinical Medicine Insights: Reproductive Health, 6*, 1–8.

Gomes, M. M., Saunders, C., & Ramalho, A. (2010). Placenta: A possible predictor of vitamin A deficiency. *British Journal of Nutrition, 103*, 1340–1344.

Gosai, S., Broadbent, R. S., Barker, D. P., Jackson, P. M., & Wheeler, B. J. (2014). Medical and midwifery attitudes towards vitamin K prophylaxis in New Zealand neonates. *Journal of Paediatrics and Child Health, 50*, 536–539.

Goyal, R., Zhang, L., Blood, A. B., Baylink, D. J., Longo, L. D., Oshiro, B., & Mata-Greenwood, E. (2014). Characterization of an animal model of pregnancy-induced vitamin D deficiency due to metabolic gene dysregulation. *American Journal of Physiology, Endocrinology and Metabolism, 306*, E256–E266.

Grabosch, S., Pennycook, J., Pakravan, A., & Mostello, D. (2013). Short bowel syndrome causing bleeding diathesis and profound vitamin deficiency in pregnancy. *Obstetrics and Gynecology, 121*, 434–436.

Grant, C. C., Stewart, A. W., Scragg, R., Milne, T., Rowden, J., Ekeroma, A., Wall, C., Mitchell, E. A., Crengle, S., Trenholme, A., Crane, J., & Camargo, C. A., Jr. (2014). Vitamin D during pregnancy and infancy and infant serum 25-hydroxyvitamin D concentration. *Pediatrics, 133*, e143–e153.

Gysler, S. M., Mulla, M. J., Stuhlman, M., Sfakianaki, A. K., Paidas, M. J., Stanwood, N. L., Gariepy, A., Brosens, J. J., Chamley, L. W., & Abrahams, V. M. (2015). Vitamin D reverses aPL-induced inflammation and LMWH-induced sFlt-1 release by human trophoblast. *American Journal of Reproductive Immunology, 73*, 242–250.

Hartmann, S., Brors, O., Bock, J., Blomhoff, R., Bausch, J., Wiegand, U. W., Hartmann, D., & Hornig, D. H. (2005). Exposure to retinyl esters, retinol, and retinoic acids in non-pregnant women following increasing single and repeated oral doses of vitamin A. *Annals of Nutrition and Metabolism, 49*, 155–164.

Hollis, B. W., Johnson, D., Hulsey, T. C., Ebeling, M., & Wagner, C. L. (2011). Vitamin D supplementation during pregnancy: Double-blind, randomized clinical trial of safety and effectiveness. *Journal of Bone and Mineral Research, 26*, 2341–2357.

Hovdenak, N., & Haram, K. (2012). Influence of mineral and vitamin supplements on pregnancy outcome. *European Journal of Obstetrics, Gynecology, and Reproductive Biology, 164*, 127–132.

Institute of Medicine Panel on Micronutrients. (2001). *Dietary Reference Intakes for Vitamin A, Vitamin K, Arsenic, Boron, Chromium, Copper, Iodine, Iron, Manganese, Molybdenum, Nickel, Silicon, Vanadium, and Zinc*. Washington, DC: National Academies Press (US). Copyright 2001 by the National Academy of Sciences. All rights reserved.

Jani, R., Palekar, S., Munipally, T., Ghugre, P., & Udipi, S. (2014). Widespread 25-hydroxyvitamin D deficiency in affluent and nonaffluent pregnant Indian women. *BioMed Research International, 2014*, 892162.

Jans, G., Guelinckx, I., Voets, W., Galjaard, S., Van Haard, P. M., Vansant, G. M., & Devlieger, R. (2014). Vitamin K1 monitoring in pregnancies after bariatric surgery: A prospective cohort study. *Surgery for Obesity and Related Diseases, 10*, 885–890.

Jishage, K., Arita, M., Igarashi, K., Iwata, T., Watanabe, M., Ogawa, M., Ueda, O., Kamada, N., Inoue, K., Arai, H., & Suzuki, H. (2001). Alpha-tocopherol transfer protein is important for the normal development of placental labyrinthine trophoblasts in mice. *Journal of Biological Chemistry, 276*, 1669–1672.

Johansen, A. M., Lie, R. T., Wilcox, A. J., Andersen, L. F., & Drevon, C. A. (2008). Maternal dietary intake of vitamin A and risk of orofacial clefts: A population-based case-control study in Norway. *American Journal of Epidemiology, 167*, 1164–1170.

Karabulut, A., Sevket, O., & Acun, A. (2011). Iron, folate and vitamin B12 levels in first trimester pregnancies in the Southwest region of Turkey. *Journal of the Turkish German Gynecology Association, 12*, 153–156.

Kasimanickam, R. K., Kasimanickam, V. R., Rodriguez, J. S., Pelzer, K. D., Sponenberg, P. D., & Thatcher, C. D. (2010). Tocopherol induced angiogenesis in placental vascular network in late pregnant ewes. *Reproductive Biology and Endocrinology, 8*, 86.

Lacroix, M., Battista, M. C., Doyon, M., Houde, G., Menard, J., Ardilouze, J. L., Hivert, M. F., & Perron, P. (2014). Lower vitamin D levels at first trimester are associated with higher risk of developing gestational diabetes mellitus. *Acta Diabetologica, 51*, 609–616.

Lane, A. S., Stallworth, J. L., Eichelberger, K. Y., & Trofatter, K. F. (2015). Vitamin K deficiency embryopathy from hyperemesis gravidarum. *Case Reports in Obstetrics and Gynecology, 2015*, 324173.

Liu, N. Q., Ouyang, Y., Bulut, Y., Lagishetty, V., Chan, S. Y., Hollis, B. W., Wagner, C., Equils, O., & Hewison, M. (2013). Dietary vitamin D restriction in pregnant female mice is associated with maternal hypertension and altered placental and fetal development. *Endocrinology, 154*, 2270–2280.

Ma, R., Gu, Y., Zhao, S., Sun, J., Groome, L. J., & Wang, Y. (2012). Expressions of vitamin D metabolic components VDBP, CYP2R1, CYP27B1, CYP24A1, and VDR in placentas from normal and preeclamptic pregnancies. *American Journal of Physiology, Endocrinology and Metabolism, 303*, E928–E935.

Mahon, P., Harvey, N., Crozier, S., Inskip, H., Robinson, S., Arden, N., Swaminathan, R., Cooper, C., & Godfrey, K. (2010). Low maternal vitamin D status and fetal bone development: Cohort study. *Journal of Bone and Mineral Research, 25*, 14–19.

Marceau, G., Gallot, D., Lemery, D., & Sapin, V. (2007). Metabolism of retinol during mammalian placental and embryonic development. *Vitamins and Hormones, 75*, 97–115.

Mason, J. B., Shrimpton, R., Saldanha, L. S., Ramakrishnan, U., Victora, C. G., Girard, A. W., Mcfarland, D. A., & Martorell, R. (2014). The first 500 days of life: Policies to support maternal nutrition. *Global Health Action, 7*, 23623.

Mcardle, H. J., Lang, C., Hayes, H., & Gambling, L. (2011). Role of the placenta in regulation of fetal iron status. *Nutrition Reviews, 69*(Suppl 1), S17–S22.

Mccarthy, M. (2014). Parents who refuse vitamin K for newborns are also likely to refuse vaccinations, Canadian study finds. *BMJ, 349*, g5283.

Miller, R. K., Hendrickx, A. G., Mills, J. L., Hummler, H., & Wiegand, U. W. (1998). Periconceptional vitamin A use: How much is teratogenic? *Reproductive Toxicology, 12*, 75–88.

Mithal, A., & Kalra, S. (2014). Vitamin D supplementation in pregnancy. *Indian Journal of Endocrinology and Metabolism, 18*, 593–596.

Muller-Schmehl, K., Beninde, J., Finckh, B., Florian, S., Dudenhausen, J. W., Brigelius-Flohe, R., & Schuelke, M. (2004). Localization of alpha-tocopherol transfer protein in trophoblast, fetal capillaries' endothelium and amnion epithelium of human term placenta. *Free Radical Research, 38*, 413–420.

Ohishi, A., Nakashima, S., Ogata, T., & Iijima, S. (2014). Early vitamin K deficiency bleeding in a neonate associated with maternal Crohn's disease. *Journal of Perinatology, 34*, 636–639.

Omeljaniuk, W. J., Socha, K., Borawska, M. H., Charkiewicz, A. E., Laudanski, T., Kulikowski, M., & Kobylec, E. (2015). Antioxidant status in women who have had a miscarriage. *Advances in Medical Sciences, 60*, 329–334.

Orendi, K., Gauster, M., Moser, G., Meiri, H., & Huppertz, B. (2010). Effects of vitamins C and E, acetylsalicylic acid and heparin on fusion, beta-hCG and PP13 expression in BeWo cells. *Placenta, 31*, 431–438.

Orzalesi, M., & Colarizi, P. (1982). Critical vitamins for low birthweight infants. *Acta Paediatrica Scandinavica - Supplement, 296*, 104–109.

Patrick, R. P., & Ames, B. N. (2015). Vitamin D and the omega-3 fatty acids control serotonin synthesis and action, part 2: relevance for ADHD, bipolar, schizophrenia, and impulsive behavior. *FASEB J, 29*(6), 2207–22.

Poston, L., Igosheva, N., Mistry, H. D., Seed, P. T., Shennan, A. H., Rana, S., Karumanchi, S. A. & Chappell, L. C. (2011). Role of oxidative stress and antioxidant supplementation in pregnancy disorders. *American Journal of Clinical Nutrition,* 94, 1980s–1985s.

Quadro, L., Hamberger, L., Gottesman, M. E., Wang, F., Colantuoni, V., Blaner, W. S., & Mendelsohn, C. L. (2005). Pathways of vitamin A delivery to the embryo: Insights from a new tunable model of embryonic vitamin A deficiency. *Endocrinology, 146*, 4479–4490.

Ross, A. C., & Gardner, E. M. (1994). The function of vitamin A in cellular growth and differentiation, and its roles during pregnancy and lactation. *Advances in Experimental Medicine and Biology, 352*, 187–200.

Rothman, K. J., Moore, L. L., Singer, M. R., Nguyen, U. S., Mannino, S., & Milunsky, A. (1995). Teratogenicity of high vitamin A intake. *New England Journal of Medicine, 333*, 1369–1373.

Schjoldager, J. G., Tveden-Nyborg, P., & Lykkesfeldt, J. (2013). Prolonged maternal vitamin C deficiency overrides preferential fetal ascorbate transport but does not influence perinatal survival in guinea pigs. *British Journal of Nutrition, 110*, 1573–1579.

Shin, J. S., Choi, M. Y., Longtine, M. S., & Nelson, D. M. (2010). Vitamin D effects on pregnancy and the placenta. *Placenta, 31*, 1027–1034.

Strom, M., Halldorsson, T. I., Hansen, S., Granstrom, C., Maslova, E., Petersen, S. B., Cohen, A. S., & Olsen, S. F. (2014). Vitamin d measured in maternal serum and offspring neurodevelopmental outcomes: A prospective study with long-term follow-up. *Annals of Nutrition and Metabolism, 64*, 254–261.

Sundrani, D., Chavan-Gautam, P., Pisal, H., Mehendale, S., & Joshi, S. (2013). Matrix metalloproteinases-2, -3 and tissue inhibitors of metalloproteinases-1, -2 in placentas from preterm pregnancies and their association with one-carbon metabolites. *Reproduction (Cambridge, England), 145*, 401–410.

Tande, D. L., Ralph, J. L., Johnson, L. K., Scheett, A. J., Hoverson, B. S., & Anderson, C. M. (2013). First trimester dietary intake, biochemical measures, and subsequent gestational hypertension among nulliparous women. *Journal of Midwifery and Women's Health, 58*, 423–430.

Veen, W. A., & Grimbergen, A. H. (1975). Vitamin E and selenium in the feed of farm animals (author's transl). *Tijdschrift voor Diergeneeskunde, 100*, 915–926.

Wei, S. Q., Audibert, F., Hidiroglou, N., Sarafin, K., Julien, P., Wu, Y., Luo, Z. C., & Fraser, W. D. (2012). Longitudinal vitamin D status in pregnancy and the risk of pre-eclampsia. *BJOG, 119*, 832–839.

Wei, S. Q., Audibert, F., Luo, Z. C., Nuyt, A. M., Masse, B., Julien, P., Fraser, W. D., & Group, M. S. (2013). Maternal plasma 25-hydroxyvitamin D levels, angiogenic factors, and pre-eclampsia. *American Journal of Obstetrics and Gynecology, 208*(390), e1–e6.

Xu, L., Lee, M., Jeyabalan, A., & Roberts, J. M. (2014). The relationship of hypovitaminosis D and IL-6 in preeclampsia. *American Journal of Obstetrics and Gynecology, 210*(149), e1–e7.

Zehnder, D., Evans, K. N., Kilby, M. D., Bulmer, J. N., Innes, B. A., Stewart, P. M., & Hewison, M. (2002). The ontogeny of 25-hydroxyvitamin D(3) 1alpha-hydroxylase expression in human placenta and decidua. *American Journal of Pathology, 161*, 105–114.

Zerfu, T. A., & Ayele, H. T. (2013). Micronutrients and pregnancy; effect of supplementation on pregnancy and pregnancy outcomes: A systematic review. *Nutrition Journal, 12*, 20.

Zingg, J. M., Meydani, M., & Azzi, A. (2012). alpha-Tocopheryl phosphate – An activated form of vitamin E important for angiogenesis and vasculogenesis? *Biofactors, 38*, 24–33.

Chapter 7
Endocrine Factors and Their Effects on Placentation

Contents

7.1 Introduction

The endocrinology of human pregnancy involves endocrine and metabolic changes that result from physiological alterations at the boundary between mother and fetus. Known as the feto-placental unit, this interface is a major site of protein and hormone production and secretion. Many of the endocrine and metabolic changes that occur during pregnancy can be directly attributed to hormonal signals originating from this unit. Maternal adaptations to hormonal changes that occur during pregnancy directly affect the development of the fetus and placenta. Gestational adaptations that take place in pregnancy include establishment of a receptive endometrium, implantation and the maintenance of early pregnancy, modification of the maternal system in order to provide adequate nutritional support for the developing fetus, and preparation for parturition and subsequent lactation. As described already, a successful human pregnancy requires cytotrophoblasts from the fetal portion of the placenta to adopt tumor-like properties. So, migration and invasion of cytotrophoblasts into the maternal endometrium are key events in human placentation. Trophoblast infiltration of maternal decidua and spiral arteries is regulated by a fine balance between the production of multiple metalloproteinases (MMP) such as MMP1–MMP16 (gelatinases, collagenases, and stromelysins), their physiological activators, and their inhibitors (TIMP1 and TIMP2) (Cohen and Bischof 2007). Many factors have been shown to regulate

A.K. Duttaroy, S. Basak, *Early Nutrition and Lifestyle Factors*,
DOI 10.1007/978-3-319-38804-5_7

invasive capacities of EVTs. Some cytokines, such as adiponectin, interleukin (IL) IL-6, IL-10, and IL-11, leukemia inhibitory factor, CCNs, epidermal growth factor, angiotensin II, and leptin, have recently been described to modulate synthesis and release of one or more of these MMPs and TIMPs (Castellucci et al. 2000; Gonzalez et al. 2001; Qiu et al. 2004). In this chapter we will describe the key endocrine/cytokine factors involved in early placentation.

7.2 Roles of Cytokines on Placentation

Cytokines play important roles during early feto-placental development and also significantly contribute to embryo elongation and attachment. In addition to different growth factors, several cytokines also influence trophoblast migration, proliferation, and invasion. As a result of a highly complex network of cytokine cellular signaling as well as cell growth or death, the feto-placental interface is an ever-changing environment throughout gestation. Many cellular processes such as angiogenesis, inflammation, and cell death are regulated by a wide variety of cytokines. To date, expression profiles for important angiogenic, inflammatory, and regulatory genes expressed at the feto-maternal unit interface throughout the pregnancy was investigated. Predominant immune interactions in the decidua between the placental trophoblast and maternal natural killer (NK) are critical. Uterine NK cells secrete different kinds of cytokines and growth factors including TNF-α, IL-10, GMCSF, IL-1, TGF-β1, CSF-1, LIF and IFN-γ. The CCNs (CYR61 (CCN1), CTGF (CCN2), and NOV (CCN3)) are crucial for angiogenesis during placental development. Expression of CCNs is described in the human placenta (Mo et al. 2002; Gellhaus et al. 2006). CCNs are strongly expressed in invasive interstitial EVTs as well as in placental endothelial cells and are significantly downregulated in preeclamptic villous placental tissue as well as in serum of preeclamptic patients (Winterhager and Gellhaus 2014). Decrease in CYR61 and NOV in PE levels could be the reason underlying the failure of uterine vascular remodeling. CTGF is expressed in the third-trimester placental villous tissue and primarily in trophoblasts, in villous endothelial cells, and, to a lesser degree, in some of the adjacent connective tissue cells within the villous core but not in first-trimester trophoblasts (Rimon et al. 2008; Waddell et al. 2011). Prokineticin-1 (PROK1)-mediated CTGF secretion in endometrial epithelial cells may act upon first-trimester trophoblast cells in a paracrine manner to promote requisite cellular steps of trophoblast conversion of maternal spiral arteries (Waddell et al. 2011). Previous study demonstrated that during PE, hypoxic conditions resulted in increased production of CTGF in trophoblast cells (Winterhager and Gellhaus 2014). CTGF has been implicated as a regulator of both physiological and pathological angiogenesis. The effects of recombinant human CTGF (hCTGF) on first-trimester extravillous trophoblast HTR8/SVneo showed increased tube formation with the concomitant increased in expression of several genes involved in angiogenesis but mostly IL-8. hCTGF-induced stimulation of IL-8 synthesis is thought to play an important role in the angiogenesis. IL-8 is a pro-inflammatory CXC chemokine frequently co-expressed with VEGF in tumors, where they are reciprocally upregulated and cooperated in angiogenesis. Trophoblasts and placental macrophages constitutively produce IL-8. Our study demonstrated that IL-8 is regulated in first-trimester placental trophoblast cells by hCTGF. IL-8 is a potent promoter of angiogenesis (Lattanzio et al. 2013). IL-8 stimulated trophoblast cell migration and invasion via stimulation of MMP2 and MMP9 levels and integrins in these cells (Lattanzio et al. 2013). The biological effects of IL-8 are mediated through the binding of IL-8 to two cell-surface G protein-coupled receptors, termed CXCR1 and CXCR2. hCTGF therefore may execute its angiogenic activity by stimulating these pro-angiogenic factors including IL-8. Though IL-8 and VEGF synergistically are reported to induce and maintain angiogenesis in endothelial cells, IL-8 can operate independently from VEGF and compensate for it in tumors (Huang et al. 2010). However, little information is available as to how IL-8 works with VEGF and other angiogenic factors in first-trimester trophoblasts. It has been suggested that endometrial IL-8 can act as a survival factor for trophoblasts. Our data are supportive of this possibility as the viable HTR8/SVneo cell numbers were increased upon treatment with IL-8. A similar influence of IL-8 on cell proliferation was shown for tumor cells (Hirota et al. 2009). A range of other chemokine receptors are also present on trophoblast, and some of their ligands promote trophoblast migration (Hanna et al. 2006; Hannan et al. 2006). Inflammatory signals associated with chemokine often lead to cellular hypoxia, which can induce angiogenesis through the upregulation of VEGF or ANGPTL4. There are considerable

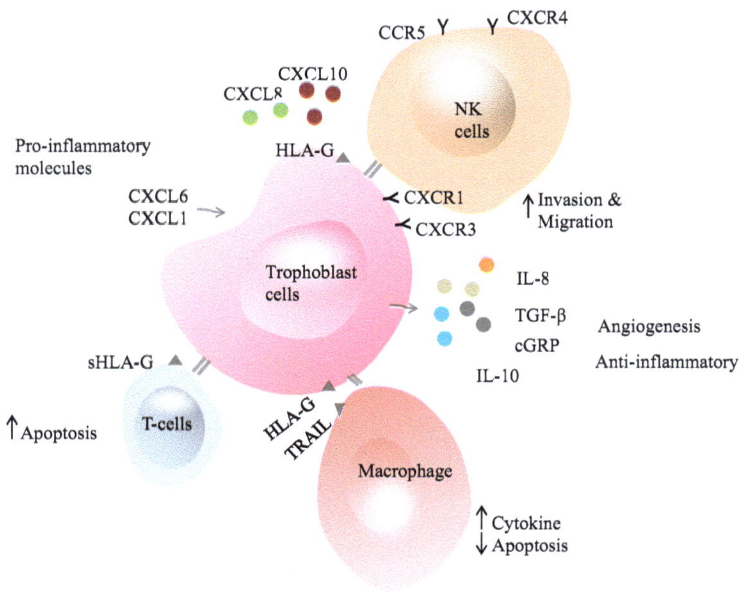

Fig. 7.1 Involvement of cytokine/chemokine sequestration during placentation

evidences to suggest that angiogenesis and chronic inflammation are co-dependent (Imhof and Aurrand-Lions 2006). Thus, a more complex interaction is likely to emerge regarding the role of chemokines in trophoblast angiogenesis. Apart from IL-8 (CXCL8), other inflammatory chemokines such as CXCL6 (IL-6) and CXCL1 (GRO-α) are implicated in tumor growth, inflammation, angiogenesis, and metastasis (Lattanzio et al. 2013; Lee et al. 2013). CTGF downregulated EDN-1, FIGF, and IFNG genes in these cells (Veillette and von Schroeder 2004; Wolf et al. 2014). An increased knowledge of the coordinated multifunctional properties of the cytokines and their signaling cascades would require further to correct multiple impaired pathways in reproductive diseases (Fig. 7.1).

7.3 Roles of Leptin and Adiponectin on Early Placental Formation

Regulation of angiogenesis by leptin has been demonstrated. Leptin is produced by placental syncytiotrophoblasts during pregnancy. Leptin protein and mRNA levels are increased in the placenta in preeclampsia (PE). Leptin stimulates tube formation in first-trimester trophoblast cells, HTR8/SVneo with concomitant increase in the expression of genes involved angiogenesis pathway. Leptin increased the expression of NRP-1. Both NRP-1 mRNA and protein are highly expressed in early trimester placenta, confirming its importance in placentation (Basak and Duttaroy 2012). In human umbilical vein endothelial cells, leptin through the binding of the membrane-bound receptors (OB-Rb) induces phosphorylation of vascular endothelial growth factor receptor 2 (VEGF-R2). Activated VEGF-R2 signals through the p38 MAPK and Akt (alpha-serine/threonine protein kinase) pathways to induce proliferation, motility, and angiogenesis. The potential relevance of these processes becomes clear in placental development and function since angiogenesis in the villi is critical for adequate fetal derived vascularization of the placenta and vascular smooth muscle growth is essential for vasomotor control of these vessels in stem villi. The systemic inflammatory response and high circulating leptin levels in pre-gravid obese women are believed to be a primer for placental inflammation during obese pregnancies. The resulting placental inflammation is reflected in the production of pro-inflammatory cytokines such as IL-6 and TNFα by activated macrophages in the placental stromal layer. This obesity-induced placental pro-inflammatory state can potentially mimic noninfectious placental villitis resulting in placental damage and altered function. Although there were no changes in placental resident immune cell populations or expression of IL-6 or TNFα, however an increases in pro-inflammatory cytokines, IL-1β, IL-8, and MCP-1 (monocyte chemotactic protein 1) in the placental villi as well as increased neutrophil populations.

Placental leptin production has shown to be in part regulated by hormonal controls. Human chorionic gonadotropin (hCG) regulates placental leptin synthesis. hCG increases both leptin mRNA and protein secretion from syncytiotrophoblast via cAMP as well as through a cross talk mechanism with p38 MAPK pathways.

hCG level rapidly increases in early pregnancy and peaks during the first trimester. These levels decline during the later part of the first trimester and are maintained at low concentrations throughout the remainder of gestation. However, as pregnancy progresses, hCG declines and the onset of pregnancy-induced maternal adipose tissue accretion likely takes over as a major contributor of systemic leptin production.

Leptin has also been shown to be regulated by 17β-estradiol (E2) in a genomic and nongenomic regulatory pathway either through nuclear estrogen receptor-α (genomic) or by an unknown membrane-bound receptor (nongenomic). During pregnancy the levels of E2 increase as a result of placental production. This increase of E2 helps to maintain pregnancy and contributes to leptin production. Leptin also regulates glucose and lipid metabolism during human development (Tessier et al. 2013).

Adiponectin has been shown to induce in vitro angiogenesis in endothelial cells via AMPK-eNOS pathway. Adiponectin plays also an important role in embryo implantation (McDonald and Wolfe 2009). This local expression of adiponectin by the endometrium, which expresses mainly AdipoR during the embryo implantation period, is important for adiponectin effects (Takemura et al. 2006). Secondly, adiponectin at the feto-maternal interface originates from maternal blood. Adiponectin is thus found in this interface very early when remodeling of spiral arteries operates. Finally, adiponectin presence at the feto-maternal barrier is relayed by the production and secretion by fetal tissues. AdipoR1 and AdipoR2 are expressed in human first-trimester placenta and also in HTR8/SVneo cells (Benaitreau et al. 2010). Adiponectin promotes trophoblast motility in a dose-independent manner in HTR8/SVneo cells. Adiponectin modulates cell migration by regulating expression of integrins such as $\alpha_5\beta_1$ and $\alpha_2\beta_1$. Adiponectin acts as a pro-invasive factor in trophoblasts. These effects were accompanied by activation of MMP2 and MMP9 and repression of TIMP2 with no changes in TIMP1 levels. Similar results were obtained in human EVTs suggesting that MMP/TIMP balance is a part of the mechanism required for adiponectin to induce trophoblast migration. To gain insights into the molecular mechanisms of adiponectin-induced trophoblast invasion, activation of the major signal transduction pathway, AMPK, was investigated in HTR8/SVneo cells (Kadowaki and Yamauchi 2005). AICAR, a specific AMPK activator, did not affect HTR8/SVneo cell invasion capacities. In human trophoblastic cell lines JEG-3 and BeWo, adiponectin antiproliferative actions seem to be mediated at least in part by the MAPK and PI3K signaling pathways (Benaitreau et al. 2009). Altered plasma level of adiponectin was reported in women with PE (Herse et al. 2009) reinforcing its possible implication in PE. EVT invasion involves proteolytic degradation of decidual and endothelial extracellular matrix (ECM) in the direction of migration followed by EVT adhesion to ECM elements and active cell migration through the degraded matrix (Mareel and Leroy 2003). When trophoblasts acquire the invasive phenotype, a downregulation of integrin $\alpha_6\beta_4$ and an upregulation of integrins $\alpha_5\beta_1$ and $\alpha_1\beta_1$ are observed suggesting that switching of integrin expression contributes to this process (Damsky et al. 1994).

Trophoblast invasion depends on a controlled program of intercellular signaling mediated by growth factors, cytokines, and hormones of both fetal and maternal origin (Bischof and Campana 2000; Cohen and Bischof 2007; Knofler et al. 2008). Insulin-like growth factor-I (IGF-1) has been shown to stimulate EVT migration and invasion.

7.4 Roles of Prostaglandins on Early Placenta Development

In general metabolites of omega-6 polyunsaturated fatty acids (LCPUFAs) are associated with increased levels of tumor angiogenesis. Eicosanoids produced from omega-6 LCPUFAs are responsible for these effects. LCPUFA when liberated by phospholipase A2 from cell plasma membrane as free fatty acids, cyclooxygenase (COX) converts the fatty acid into prostaglandins. For example, arachidonic acid, 20:4$n-6$ (ARA) is converted to prostaglandin (PG) E_2, PGD_2, PGI_2, and thromboxane A_2. Studies demonstrated a link between prostaglandins and cancer in particular with the E series prostaglandins. COX-2 inhibitors were shown to significantly reduce tumor formation in animal models. COX-2 is upregulated in most human cancers and PGE2 is produced in large amounts in tumors. PGE_2 was shown experimentally to induce the production of pro-angiogenic factors in many cell types. A stimulatory effect of nitric oxide on COX-2 activity in human colorectal cancers is observed, and this interaction is likely to yield a cooperative effect in promoting angiogenesis through a PGE_2 increase in VEGF production. Several animal studies showed that omega-3 fatty acid-enriched diets have inhibitory effects on COX-2 and prostaglandin production in both plasma and experimentally induced tumors. Tumor incidence was significantly decreased in the $n-3$ fatty acids group compared to the $n-6$ group, and levels of COX-1 and COX-2 were significantly reduced in the $n-3$ fatty acid group. EPA and DHA reduced VEGF, COX-2 expression, and PGE_2 levels in HT-29 cells cultured in vitro. EPA and DHA also inhibited ERK-1 and ERK-2 phosphorylation and HIF-1alpha protein overexpression, critical steps in the PGE_2-induced signaling pathway leading to the augmented expression of VEGF in colon cancer cells. Prostaglandins are involved in the implantation of embryos as well as the proper development of early placentation (Qian et al. 2011; Kaczynski et al. 2016; Vilella et al. 2013). Abnormal prostaglandin synthesis is shown to be responsible for pregnancy loss in humans (Achache et al. 2010). However, the identity of responsible prostaglandins is not yet known as general inhibitors of prostaglandin synthesis were used. Although VEGF is the main player in angiogenesis, recent data indicate that the angiogenic process is more complex and may involve many other factors. In fact, several fatty acids stimulate angiogenesis possibly via different mechanisms including prostaglandin productions (Basak et al. 2013). In fact PGE_2 is involved in angiogenesis by acting indirectly on a variety of cell types by stimulating the angiogenic factors. Indeed, PGE_2 increased production of VEGF, bFGF, or CXCL1 that, in turn, act on target endothelial cells to promote the angiogenesis (Spencer et al. 2009). PGE_2 is produced from ARA by COX-2. There are ample evidences to implicate COX-2 as an angiogenesis mediator. COX-2 is an inducible enzyme which plays a critical role in multiple pathophysiological processes including angiogenesis. It has been shown that overexpression of COX-2 is significantly correlated to invasiveness, prognosis, and survival in some cancers. Inhibition of COX-2 with selective COX-2 inhibitors effectively prevents inflammation, proliferation, and angiogenesis and induces apoptosis in human cells. Small animal studies demonstrated that omega-3 fatty acids had antiproliferative and anti-angiogenic effects (Spencer et al. 2009). EPA and DHA inhibit production of many important angiogenic factors VEGF, platelet-derived growth factor (PDGF), COX-2, PGE2, nitric oxide, NFkB, MMPs, and beta-catenin (Spencer et al. 2009). However, the exact mechanisms as to how omega 3 fatty acids impart anti-angiogenesis are not yet fully known. Contrary to this, DHA and other fatty acids including conjugated linoleic acid stimulate angiogenesis in placental first-trimester cells (Basak and Duttaroy 2013) as shown in Fig. 7.2. In addition to the stimulation of expression of major angiogenic factors

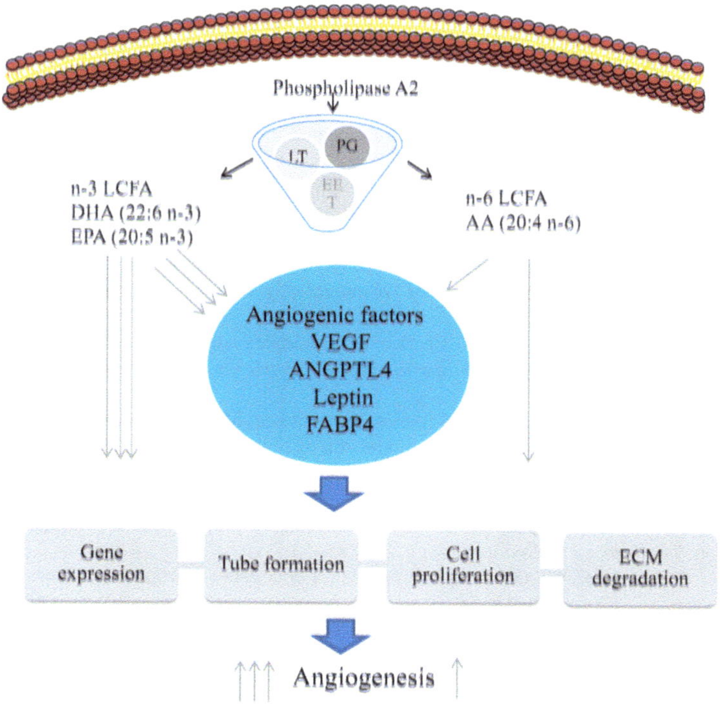

Fig. 7.2 Prostaglandin-derived long-chain fatty acids induce pro-angiogenic stimulation in the first-trimester placenta. *LCFAs* long-chain fatty acids, *LTs* leukotrienes, *PGs*, prostaglandins, *EETs* epoxyeicosatrienoic acids, *ECM* extracellular matrix, *AA* arachidonic acid, *DHA* docosahexaenoic acid, *EPA* eicosapentaenoic acid, *VEGF* vascular endothelial growth factor, *FABP* fatty acid-binding protein, *ANGPTL4* angiopoietin-4

such as VEGF and ANGPTL4, fatty acids also stimulate expression of intracellular fatty acid-binding proteins (FABPs) FABP-4 and FABP-3 which are known to directly modulate angiogenesis (Basak and Duttaroy 2013).

7.5 Summary

Apart from major angiogenic factors such as VEGF, ANGPTLs, the endocrine factors such as cytokines are also involved in angiogenesis in the placenta. CCNs, leptin, and interleukins are involved in trophoblast cell growth and proliferation and angiogenesis. Dietary fatty acids also affect the process via modulating the levels of VEGF, PDGF, COX-2, and PGE2. Usually $n - 3$ fatty acids inhibit and $n - 6$ fatty acids stimulate angiogenesis in tumors and other cells. However, in placental trophoblasts, all types of fatty acids ($c > 16$) stimulate tube formation (as a measure of angiogenesis), although the mechanisms are not known yet.

References

Achache, H., Tsafrir, A., Prus, D., Reich, R., & Revel, A. (2010). Defective endometrial prostaglandin synthesis identified in patients with repeated implantation failure undergoing in vitro fertilization. *Fertility and Sterility, 94*, 1271–1278.

Basak, S., Das, M. K., & Duttaroy, A. K. (2013). Fatty acid-induced angiogenesis in first trimester placental trophoblast cells: Possible roles of cellular fatty acid-binding proteins. *Life Sciences, 93*, 755–762.

Basak, S., & Duttaroy, A. K. (2012). Leptin induces tube formation in first-trimester extravillous trophoblast cells. *European Journal of Obstetrics, Gynecology, and Reproductive Biology, 164*, 24–29.

Basak, S., & Duttaroy, A. K. (2013). Effects of fatty acids on angiogenic activity in the placental extravillious trophoblast cells. *Prostaglandins, Leukotrienes, and Essential Fatty Acids, 88*, 155–162.

Benaitreau, D., Dieudonne, M. N., DOS Santos, E., Leneveu, M. C., Mazancourt, P., & Pecquery, R. (2009). Antiproliferative effects of adiponectin on human trophoblastic cell lines JEG-3 and BeWo. *Biology of Reproduction, 80*, 1107–1114.

Benaitreau, D., dos Santos, E., Leneveu, M. C., Alfaidy, N., Feige, J. J., de Mazancourt, P., Pecquery, R., & Dieudonne, M. N. (2010). Effects of adiponectin on human trophoblast invasion. *The Journal of Endocrinology, 207*, 45–53.

Bischof, P., & Campana, A. (2000). A putative role for oncogenes in trophoblast invasion? *Human Reproduction, 15*(Suppl 6), 51–58.

Castellucci, M., DE Matteis, R., Meisser, A., Cancello, R., Monsurro, V., Islami, D., Sarzani, R., Marzioni, D., Cinti, S., & Bischof, P. (2000). Leptin modulates extracellular matrix molecules and metalloproteinases: Possible implications for trophoblast invasion. *Molecular Human Reproduction, 6*, 951–958.

Cohen, M., & Bischof, P. (2007). Factors regulating trophoblast invasion. *Gynecologic and Obstetric Investigation, 64*, 126–130.

Damsky, C. H., Librach, C., Lim, K. H., Fitzgerald, M. L., Mcmaster, M. T., Janatpour, M., Zhou, Y., Logan, S. K., & Fisher, S. J. (1994). Integrin switching regulates normal trophoblast invasion. *Development, 120*, 3657–3666.

Gellhaus, A., Schmidt, M., Dunk, C., Lye, S. J., Kimmig, R., & Winterhager, E. (2006). Decreased expression of the angiogenic regulators CYR61 (CCN1) and NOV (CCN3) in human placenta is associated with pre-eclampsia. *Molecular Human Reproduction, 12*, 389–399.

Gonzalez, R. R., Devoto, L., Campana, A., & Bischof, P. (2001). Effects of leptin, interleukin-1alpha, interleukin-6, and transforming growth factor-beta on markers of trophoblast invasive phenotype: Integrins and metalloproteinases. *Endocrine, 15*, 157–164.

Hanna, J., Goldman-Wohl, D., Hamani, Y., Avraham, I., Greenfield, C., Natanson-Yaron, S., Prus, D., Cohen-Daniel, L., Arnon, T. I., Manaster, I., Gazit, R., Yutkin, V., Benharroch, D., Porgador, A., Keshet, E., Yagel, S., & Mandelboim, O. (2006). Decidual NK cells regulate key developmental processes at the human fetal-maternal interface. *Nature Medicine, 12*, 1065–1074.

Hannan, N. J., Jones, R. L., White, C. A., & Salamonsen, L. A. (2006). The chemokines, CX3CL1, CCL14, and CCL4, promote human trophoblast migration at the feto-maternal interface. *Biology of Reproduction, 74*, 896–904.

Herse, F., Bai, Y., Staff, A. C., Yong-Meid, J., Dechend, R., & Zhou, R. (2009). Circulating and uteroplacental adipocytokine concentrations in preeclampsia. *Reproductive Sciences, 16*, 584–590.

Hirota, Y., Osuga, Y., Hasegawa, A., Kodama, A., Tajima, T., Hamasaki, K., Koga, K., Yoshino, O., Hirata, T., Harada, M., Takemura, Y., Yano, T., Tsutsumi, O., & Taketani, Y. (2009). Interleukin (IL)-1beta stimulates migration and survival of first-trimester villous cytotrophoblast cells through endometrial epithelial cell-derived IL-8. *Endocrinology, 150*, 350–356.

Huang, D., Ding, Y., Zhou, M., Rini, B. I., Petillo, D., Qian, C. N., Kahnoski, R., Futreal, P. A., Furge, K. A., & Teh, B. T. (2010). Interleukin-8 mediates resistance to antiangiogenic agent sunitinib in renal cell carcinoma. *Cancer Research, 70*, 1063–1071.

Imhof, B. A., & Aurrand-Lions, M. (2006). Angiogenesis and inflammation face off. *Nature Medicine, 12*, 171–172.

Kaczynski, P., Kowalewski, M., & Waclawik, A. (2016). Prostaglandin F2alpha promotes angiogenesis and embryo-maternal interactions during implantation. *Reproduction, 151*(5), 539–52.

Kadowaki, T., & Yamauchi, T. (2005). Adiponectin and adiponectin receptors. *Endocrine Reviews, 26*, 439–451.

Knofler, M., Simmons, D. G., Lash, G. E., Harris, L. K., & Armant, D. R. (2008). Regulation of trophoblast invasion – A workshop report. *Placenta, 29*(Suppl A), S26–S28.

Lattanzio, L., Tonissi, F., Torta, I., Gianello, L., Russi, E., Milano, G., Merlano, M., & Lo Nigro, C. (2013). Role of IL-8 induced angiogenesis in uveal melanoma. *Investigational New Drugs, 31*, 1107–1114.

Lee, W. J., Lan, K. H., Chou, C. T., Yi, Y. C., Chen, W. C., Pan, H. C., Peng, Y. C., Wang, K. B., Chen, Y. C., Chao, T. H., Tien, H. R., Sheu, W. H., & Sheu, M. L. (2013). Tpl2 inhibitors thwart endothelial cell function in angiogenesis and peritoneal dissemination. *Neoplasia, 15*, 1036–1048.

Mareel, M., & Leroy, A. (2003). Clinical, cellular, and molecular aspects of cancer invasion. *Physiological Reviews, 83*, 337–376.

McDonald, E. A., & Wolfe, M. W. (2009). Adiponectin attenuation of endocrine function within human term trophoblast cells. *Endocrinology, 150*, 4358–4365.

Mo, F. E., Muntean, A. G., Chen, C. C., Stolz, D. B., Watkins, S. C., & Lau, L. F. (2002). CYR61 (CCN1) is essential for placental development and vascular integrity. *Molecular and Cellular Biology, 22*, 8709–8720.

Qian, X., Zhang, J., & Liu, J. (2011). Tumor-secreted PGE2 inhibits CCL5 production in activated macrophages through cAMP/PKA signaling pathway. *Journal of Biological Chemistry, 286*, 2111–2120.

Qiu, Q., Yang, M., Tsang, B. K., & Gruslin, A. (2004). EGF-induced trophoblast secretion of MMP-9 and TIMP-1 involves activation of both PI3K and MAPK signalling pathways. *Reproduction, 128*, 355–363.

Rimon, E., Chen, B., Shanks, A. L., Nelson, D. M., & Sadovsky, Y. (2008). Hypoxia in human trophoblasts stimulates the expression and secretion of connective tissue growth factor. *Endocrinology, 149*, 2952–2958.

Spencer, L., Mann, C., Metcalfe, M., Webb, M., Pollard, C., Spencer, D., Berry, D., Steward, W., & Dennison, A. (2009). The effect of omega-3 FAs on tumour angiogenesis and their therapeutic potential. *European Journal of Cancer, 45*, 2077–2086.

Takemura, Y., Osuga, Y., Yamauchi, T., Kobayashi, M., Harada, M., Hirata, T., Morimoto, C., Hirota, Y., Yoshino, O., Koga, K., Yano, T., Kadowaki, T., & Taketani, Y. (2006). Expression of adiponectin receptors and its possible implication in the human endometrium. *Endocrinology, 147*, 3203–3210.

Tessier, D. R., Ferraro, Z. M., & Gruslin, A. (2013). Role of leptin in pregnancy: Consequences of maternal obesity. *Placenta, 34*, 205–211.

Veillette, C. J., & von Schroeder, H. P. (2004). Endothelin-1 down-regulates the expression of vascular endothelial growth factor-A associated with osteoprogenitor proliferation and differentiation. *Bone, 34*, 288–296.

Vilella, F., Ramirez, L., Berlanga, O., Martinez, S., Alama, P., Meseguer, M., Pellicer, A., & Simon, C. (2013). PGE2 and PGF2alpha concentrations in human endometrial fluid as biomarkers for embryonic implantation. *Journal of Clinical Endocrinology and Metabolism, 98*, 4123–4132.

Waddell, J. M., Evans, J., Jabbour, H. N., & Denison, F. C. (2011). CTGF expression is up-regulated by PROK1 in early pregnancy and influences HTR-8/Svneo cell adhesion and network formation. *Human Reproduction, 26*, 67–75.

Winterhager, E., & Gellhaus, A. (2014). The role of the CCN family of proteins in female reproduction. *Cellular and Molecular Life Sciences, 71*(12), 2299–2311.

Wolf, D., Tseng, N., Seedorf, G., Roe, G., Abman, S. H., & Gien, J. (2014). Endothelin-1 decreases endothelial PPARgamma signaling and impairs angiogenesis after chronic intrauterine pulmonary hypertension. *American Journal of Physiology – Lung Cellular and Molecular Physiology, 306*, L361–L371.

Chapter 8
Maternal Lifestyle Factors and Placentation

Contents

8.1 Introduction

Developing nations carry the burden of maternal malnutrition and nutrient deficiencies that ultimately contribute to adverse fetus outcome and low birth weight (LBW)-related issues. In developed nation, undernutrition is often restricted to women with eating disorder or lifestyle (smoking/alcohol) habit that increases risk for LBW. In both the cases, placenta is exposed to certain stresses that may affect the placentation. In addition to nutrition, there are several lifestyle factors, such as cigarette smoking, environmental pollution are a global threat that includes pregnant women particularly in developed countries where 15–30 % of pregnant women smoke during pregnancy. It is important to know the effect of maternal smoking, alcohol intake, and exposure to the environmental pollutants on placentation, since pregnant women if exposed to these factors during preconception to gestational life may have an adverse consequence on pregnancy outcome. Impaired

A.K. Duttaroy, S. Basak, *Early Nutrition and Lifestyle Factors*,
DOI 10.1007/978-3-319-38804-5_8

placentation that ultimately influences fetus outcome is the center of many developmental disorders and contributes as an independent risk factor for the adult chronic diseases. This chapter will review the impact of maternal lifestyle behavior such as alcohol intake, maternal smoking, and environmental toxins on placental blood flow, vasculature, and angiogenesis of the placenta that includes the process of trophoblast growth and development trajectory linked with placentation that ultimately determines fetus growth and birth outcome. Specific effect of persistent organic pollutants (POPs) and its metabolites, nicotine, aryl hydrocarbon receptor (AHR) and their ligands on trophoblast metabolism, gene expression, nutrient uptake, and transport of the placenta and trophoblast at early and late gestation is described in this chapter.

8.2 Impact of Maternal Smoking on the Placental Trophoblast Development Mediated by Aryl Hydrocarbon Receptor

Use of tobacco is viewed as an important threat to the health of pregnant women and their children. Prevalence of current tobacco use (smoked and smokeless) estimated in pregnant women ranges from 0 to 15 % in high-income countries, including the USA (12×3 %) and the UK (36×0 %) (Caleyachetty et al. 2014). Invariably, smoking in pregnancy increases the risk of pregnancy complications such as spontaneous abortion, fetal growth restriction, preterm premature rupture of membranes, placenta previa, placental abruption, and preterm delivery. Active smoker subjects carry the risk of failed pregnancy. In contrast, the risk of PE is markedly reduced among smokers. However, the evidence of smoking on regulation of endothelial dysfunction in pregnancy is yet to be established. A comprehensive systematic review of the scientific literature on the impact of cigarette smoking and smoke constituents on the different stages of reproductive function, including epidemiological, clinical, and experimental studies is reported recently (Dechanet et al. 2011).

In the placenta, when smoking is associated from early in pregnancy, a thickening of the trophoblastic basement membrane, an increase in collagen content of the villous mesenchyme, and a decrease in vascularization have been reported. Maternal smoking dysregulates trophoblast growth and gene expression by modulating cellular oxygen tension. This leads to a reduction of fetus weight and less fat mass with altered protein metabolism. Polycyclic aromatic hydrocarbons (PAHs) are an important constituent of cigarette smoke, and the effects of acute exposure to these chemicals during placentation are not known. Several PAHs are ligands for the aryl hydrocarbon receptor (AhR). AhR is constitutively expressed protein with its increased expression in lung, placenta, spleen, and ovary. AhR is the target for a series of environmental contaminants generated through incineration such as coal, tar, and wood burning and, oil seeps, chimney soot, and smoke exhaust (Stejskalova et al. 2011). AhR regulates the expression of genes of cytochrome P450 pathway as a xenobiotic response. Cigarette smoke is known to contain large amounts of AHR agonists (Kitamura and Kasai 2007). Thus, pregnant women are likely to be exposed to various exogenous AhR ligands knowingly through smoke and pollution as well as unknowingly through the diets. Excess exposure of these ligands through environment may contribute to a nongenetic modulating factor during pregnancy. AhR and aryl hydrocarbon nuclear translocator (ARNT) mRNA are expressed in first trimester and term human placental tissues (Hakkola et al. 1997). PAHs have been shown to cross the placenta, form DNA adducts on both the maternal and the fetal sides and in trophoblast cells, and affect the placental architecture (Detmar et al. 2008). Importantly, elevated CYP1A1 activity through activated AhR in placentas of women smokers has been associated with preterm birth; intrauterine growth retardation; structural abnormalities; fetal death or placenta abruption; and risk of low birth weight, low birth length, and low head circumference. Chronic exposure to PAHs in mice before gestation results in compromised maternal and fetal vascularizations and altered labyrinthine architecture localized in placental syncytiotrophoblasts both in first trimester and full-term placenta (Detmar et al. 2008). CYP1A1 is the only placental xenobiotic-metabolizing enzyme of the cytochrome P450 superfamily series, for which significant expression and catalytic activity have been conclusively demonstrated in placental trophoblasts throughout human pregnancy (Myllynen et al. 2009). AhR and ARNT proteins are expressed in first trimester placenta, whereas AhR is expressed low in placenta as compared with other tissues. Localization and expression of human AhR and ARNT in the first trimester and in the syncytiotrophoblast of term placentas indicate their role in facilitating placental xenobiotic response. Among the different CYP1 genes tested, only CYP1A1 mRNA is significantly upregulated in isolated trophoblasts after exposure to AhR ligands which is similar to placental induction of CYP1A1 in smoking women (Bruchova et al. 2010). CYP1A2 is mainly a hepatic enzyme; its mRNA has been detected in the first trimester placentas but not in full-term placentas. UGT1A subfamily mRNA was found in first trimester placentas but not in term placentas, and UGTA1 activity has been reported to be more significant in smoking women. Importantly, elevated CYP1A1 placental activity in smoking women has been associated with adverse birth outcomes (such as premature birth, intrauterine growth retardation, and structural abnormality). CYP1B1 mRNA is constitutively expressed at a low mRNA level in first trimester and full-term placentas (Hakkola et al. 1997). CYP1B1 mRNA is not induced in the placenta by maternal smoking. In contrast, another study (Bruchova et al. 2010) reports that smoking during human pregnancy induces expression of CYP1B1 mRNA level in the placenta.

Smoking active compound like PAHs can accumulate in adipose and mammary tissues before smoking cessation and can be released into the blood stream during pregnancy. More than a year after smoking cessation, PAH-DNA adducts in human blood samples are reduced by only 50 % compared with levels during smoking. PAHs have also been shown to accumulate in the placentas of non and ex-smokers in heavily polluted areas (Gladen et al. 2000). Feto-placental vascular resistance appears to be chronically elevated by smoking (Morrow et al. 1988; Kimya et al. 1998; Albuquerque et al. 2004), and this is anticipated due to reduced feto-placental blood flow. A sustained increase in resistance of the blood flow suggests that smoking causes alteration in vascularization of the blood vessel. Smoking is associated with decreases in feto-placental capillary surface area, volume, and length in human placentas (Larsen et al. 2002). Pre-pregnancy exposure to PAH anticipated with low blood flow can contribute approximately 23 % reduction in fetal weight (Rennie et al. 2011). In contrast, the data predicted that women who smoke throughout the pregnancy are likely to have a 33 % less risk of developing PE. The effect of smoking on protecting preeclampsia is multifactorial that includes several players such as VEGF, hypoxia, AhR, and others. The active compound of cigarette, cotinine, and other substances seem to trigger the activation of AhR, a "dioxin-like potential" compound. These substances upon binding to AhR translocate to the nucleus where it complexes with the aryl hydrocarbon receptor nuclear transporter (ARNT) protein. The ligand-activated AhR/ARNT heterodimers bind to specific enhancer sequences located upstream in the promoter of aromatic hydrocarbon-inducible genes. Several possible mechanisms are suggested for protective effects of cigarette smoke on PE. High protein levels of AhR from normal, mild preeclampsia, and severe preeclampsia women indicate its physiological role in maintaining the placentation and the development of the fetus. Since activation of AhR significantly enhanced the invasion of cancer cells, therefore, high levels of dioxin-like potential compounds in cigarette smoke may increase trophoblast invasion (common pathway for cancer metastasis) and spiral arteriole remodeling by activating AhR–XRE pathways (Wang et al. 2011).

8.3 Effect of Smoking on the Placental Trophoblast Development by Angiogenic Factors

Interestingly, a considerable amount of preterm birth (15 % of the total) is derived from preeclampsia where major risk factors for this disease are primiparity, previous preeclampsia, and chronic hypertension. Numerous therapies that included salt restriction, zinc, magnesium, fish oil, calcium, and vitamins were unable to demonstrate significant benefit in the prevention of preeclampsia. Early delivery of the fetus may reduce maternal morbidity at the cost of fetal interests in earlier gestations. Inadequacy of placental angiogenesis due to excess soluble fms-like tyrosine kinase-1 (sFlt-1), a vascular endothelial growth factor receptor bound available VEGF to augment angiogenesis in the placenta is commonly observed. Paradoxically, it has been noted over the years that the risk of preeclampsia is lower in smokers than in nonsmokers (Lain et al. 1999; Hammoud et al. 2005). It is notable that, in men and nonpregnant women, the levels of sFlt-1 in smokers are one-half that of nonsmokers; similarly, in pregnancy, serum levels of sFlt-1 are lower in smokers compared with nonsmokers (Belgore et al. 2000). Cigarette smoke extract was estimated based on the secretion of sFlt-1 in the term placental tissue (Mehendale et al. 2007). Aqueous extract of cigarette smoke was used in a unique way to expose the cells that somewhat mimics the in vivo situation. Exposure to cigarette smoke extract reduced cell soluble secretion of sFlt-1 level in a dose-dependent manner. It seems that exposure of placental villous explants to cigarette smoke extract turns the placental villous cells in a pro-angiogenic state with reduced level of sFlt-1 secretion and relative abundance of PlGF. This is the reverse of changes of sFlt-1/PIGF ratio that are seen in preeclampsia and may explain the reduction of preeclampsia in smokers.

Meta-analysis of 34 studies conclusively states that there is a dose-dependent reduction in the risk of PE by maternal cigarette smoking irrespective of sample size and geography (Conde-Agudelo et al. 1999). The similar protective effect of maternal cigarette smoking was reported from analysis of German perinatal database of 157,857 singleton pregnancies (Hammoud et al. 2005). Inverse correlation between the rate of preeclampsia and the number of cigarettes smoked per day was observed in 23 % of pregnant smokers of the database. In another study, maternal tobacco exposure was quantified and classified as nonexposed (<50 ng/mL), light exposure (50–300 ng/mL), and heavy exposure (>300 ng/mL) by measuring the urinary cotinine level (Lain et al. 1999). Overall, odds ratio for exposed vs. nonexposed subjects for the development of preeclampsia was 0.31 (95 % CI, 0.12–0.79). Some studies argued that the protective effect of smoke exposure is predominant in ever smokers or in women who have recently ceased smoking, from the development of gestational hypertension but not preeclampsia (Zhang et al. 1999). The first trimester placental villus explants from patients who were exposed to passive or active smoke were compared with the expression of von Hippel Landau protein, a protein that suppresses angiogenesis (Genbacev et al. 1995). Author suggests that in vivo smoke exposure downregulates this protein, enhances angiogenesis, and may improve placental invasion. However, exposure of the explants villous cells to nicotine demonstrated the opposite effect which leads to reduce angiogenesis and hence poor placental invasion as seen in PE. The protective effect of smoking in PE could be carbon monoxide produced during smoking (Bainbridge et al. 2005). Protective effect of smoking was also observed in previous smokers who stopped smoking in early gestation and those no longer exposed to carbon monoxide continuously (Conde-Agudelo et al. 1999). All these reports support the idea of protecting PE by smoking and it seems that the components of cigarette smoke may produce the protective effects in the pathogenesis of PE.

VEGF, VEGF receptor-1 (VEGFR-1), and VEGFR-2 are essential for maintenance of pregnancy. The soluble secreted form of VEGFR-1 (sVEGFR-1) binds to VEGF and considered as a potent negative regulator of VEGF activity in vivo. Preeclamptic women have increased levels of sVEGFR-1 and thus supports the hypothesis of inadequate VEGF-mediated angiogenesis in PE. Cigarette smoking was shown to reduce circulating sVEGFR-1 level in nonpregnant women and in men (Schmidt-Lucke et al. 2005; Belgore et al. 2000). In vitro, nicotine and its major metabolite, cotinine, upregulate cellular VEGF and VEGFR-1 gene expression. VEGF positive immunostaining in pulmonary arteries and bronchial mucosa of smokers was reported. Under hypoxia, secretion of sVEGFR1 is decreased in term placenta (Mehendale et al. 2007). Since smoking reduces the incidence of preeclampsia and affects the expression of VEGFR-1 and as VEGFR-1 expression increases in preeclampsia, several studies investigated the protective mechanism of smoking, if any, on the VEGF and its receptors, VEGFR-1, in pregnant women.

The relationships between smoking and VEGF in placenta are complex and data are inconsistent. Maternal plasma level of sVEGFR-1 in the third trimester of smoker (median 1088, range 834–1362 ng/L, $n = 20$) and nonsmoker (728, 719–1336 ng/L, $n = 19$) women was higher than its level in the second trimester (smokers: 374, 291–683 ng/L, $n = 6$, $p > 0.05$; nonsmokers: 375, 290–667 ng/L, $n = 22$, $p < 0.001$) (Kamarainen et al. 2009). No difference in the level of sVEGFR-1 was observed between smokers and nonsmokers. The circulating maternal VEGFR-1 concentrations are not affected by maternal smoking, and thus, the protective effect of smoking on PE, if any, may not be mediated by VEGFR-1. The study differs from those of Powers et al., who demonstrated that smokers have lower serum sVEGFR-1 concentrations during the second and third trimesters than nonsmokers (Powers et al. 2005). First trimester villous tissue samples ($n = 31$) were analyzed for placental expression of genes related to angiogenesis and oxidative stress pathway. Placental expression of VEGFA was significantly higher in smoking women at 10–11 weeks of gestation, not 6–7 weeks, compared with nonsmoking women at the same gestational age. There was no difference in the expression of fms-like tyrosine kinase (Flt-1), soluble endoglin (sEng), placental growth factor (PlGF), hemoxygenase-1 (HMOX-1), and superoxide dismutase (SOD) related to smoking status. Author suggested that increased VEGF expression in the early stages of pregnancy of smoking women may contribute factor to decrease the risk of developing preeclampsia (Shinjo et al. 2014). Maternal smoking made a direct toxic effect on the fetal cells or an indirect effect through damage and/or functional disturbances of the placenta (Jauniaux and Burton 2007). During epithelialization, cytotrophoblast stem cells either fuse to form the syncytium or aggregate to form cell columns that adhere to and invade in the uterus. Importantly, chorionic villi from early gestation placentas of mothers who smoke showed a marked reduction in cell columns. Exposing early gestation chorionic villi from nonsmoking women to nicotine, inhibited subsequent cell column formation in vitro. In vitro, nicotine inhibited normal first trimester cytotrophoblast invasion, apparently by reducing the ability of treated cells to synthesize and activate type IV collagenase, an important mediator of invasion in vitro (Genbacev et al. 1995). This very early study on the first trimester pregnancy suggests that nicotine may contribute to the growth retardation observed in fetuses of mothers who smoke during pregnancy.

8.4 Protection of Preeclampsia by Smoking: Role of Adrenomedullin and Carbon Monoxides on Trophoblast Invasion?

Cigarette smoke-dependent tumor growth was reversed with adrenomedullin (AM) receptor antagonist co-treatment. This work first postulates that cigarette smoke extract (CSE) enhances production of AM in the cell media that ultimately affect the cellular growth (Lenhart and Caron 2012). Murine placenta that lacks of AM gene typically shows pathological feature of preeclampsia. The fetal labyrinth vessels within the placenta of these AM knockout mice demonstrate reduced branching and less number of decidual natural killer (NK) cells with a concomitant decrease in the remodeling of maternal spiral arterioles (Li et al. 2013). This study demonstrates the importance of AM for normal placental development. Cigarette smoke extract treatment of first trimester trophoblast cells, HTR-8/SVneo, causes increased synthesis and secretion of AM, and this treatment leads to improved cellular viability and migration that are dependent on CSE-mediated cellular AM production (Kraus et al. 2014). AM has multiple functions including angiogenic, anti-inflammatory, and antimicrobial activities. AM is highly expressed in reproductive tissues, but its role in the pathogenesis of preeclampsia is not clear. AM expression was downregulated in the placentas of women with PE compared to normal controls (Gratton et al. 2003). Recently, AM has been shown to increase the invasiveness of trophoblast cells in the cellular model of invasion and that this induction was reversed with the use of its inhibitor (Kraus et al. 2014). These data raise the speculation that altered AM levels in the placenta may contribute to the pathogenesis of the PE.

The consensus from different studies has now concluded that smoking during pregnancy reduces the risk of developing PE by approximately 50 %. In order to study this phenomenon, appropriate models and doses of CSE to study the effects of cigarette smoking on trophoblast function are crucial. CSE dilution of 1 % estimates the nicotine exposure of a heavy smoker (1–2 packs per day leads to nicotine blood levels of ~200 ng/mL). The smoke generated from the cigarette in vivo does not contact any tissues directly. Rather, the phenomenon induces numerous soluble components to secrete within the systems that actually regulate the several pathophysiological responses. Not only the dose of cigarette smoking, there are several other factors such as the duration (first trimester vs. term), tenure (ever smoker vs. seasonal), and type of tobacco (smoke vs. smokeless) that contribute differentially to the protection from preeclampsia. Swedish birth register on all singleton births during the years 1999–2006 in Sweden ($n = 612{,}712$) evaluated preeclampsia risk in smokers and noticed that the risk reduction did not cover women who smoked in the first trimester and later quit by the onset of the third trimester compared to those who continued the smoking habit throughout pregnancy provided greatest protection (Wikstrom et al. 2010) The smokeless tobacco (i.e., snuff) could not offer any protection from preeclampsia. Tobacco combustion products rather than nicotine probably contain the protective ingredients against preeclampsia in cigarette smoke. Interestingly, nicotine treatment of trophoblast cells alone failed to affect cell viability, migration, or AM production, while CSE treatment enhanced cellular viability and migration, both of which were dependent on AM (Kraus et al. 2014). All these findings suggest that smoking throughout pregnancy is required to maintain protection and that nicotine products do not provide similar protection.

The smokers express more AM in term placenta tissue than the same tissue from nonsmokers (Kraus et al. 2014). Parallel to this finding, the level of placental AM production was found to be lower in preeclamptic women compared to normal mother. The plasma level of AM production rises over the course of gestation in normal pregnancies, but there are conflicting data about plasma levels of AM in preeclamptic pregnancies. Based on the level of AM during pregnancy, two theories are postulated for the role of AM in the protection of preeclampsia. Increased level of AM could be the compensatory vasodilator response attempting to counteract the vasospasm at the root of preeclamptic hypertension (Gratton et al. 2003). In contrast, lower circulatory AM levels obey the pathogenesis of the disease, since decreased vasodilation is a key contributing factor with preeclamptic phenotype (Dikensoy et al. 2009). This study demonstrated that treatment of a first trimester cell line with CSE increases cell invasion through an AM-mediated process and that placental AM levels are increased among smokers at term compared to non-smokers. It seems that cigarette smoking enhances placental trophoblast invasion by increasing cellular production of AM, which may improve placental function by improving spiral arteriole remodeling, thereby decreasing placental release of antiangiogenic factors, which have been implicated in preeclampsia. Future work with CSE treatment of trophoblast cells with AM homolog on trophoblast cellular invasion will be required to strengthen the hypothesis. Blocking the function of AM by silencing its receptor would be required to confirm its specificity in the trophoblast. AM expression and localization in the first trimester placenta and its interaction with hypoxia-mediating factors need to be studied to understand its protection on trophobalst invasion in vivo. Taken together, it is suggested that protection against the development of preeclampsia may be partially mediated by increased placental production of AM.

Now it is clear that cigarette smoking combustion product(s) is the possible mediator of the protection against the PE. Nitric oxide (NO) is known to promote vascular relaxation, and angiogenesis regulation is also produced during smoking. However, decreased circulatory level of NO metabolites in smokers ruled out its role in protecting preeclampsia in smokers. Recent data suggest that carbon monoxide (CO) may be the mediator of this protection process (Karumanchi and Levine 2010). CO is a known vascular protective agent in a number of vascular disorders, such as ischemia–reperfusion injuries. Apart from its presence as environmental pollutant, CO is produced endogenously during cellular metabolism via the hemoxygenase (HO) system. The observation provides a first-hand clue on the mechanism of smoker's protection mediated by CO where CO-releasing compounds induced lower secretion of sFlt1 and sEng in placental tissue culture and endothelial cells. It is possible that cigarette smoking directly upregulates HO expression in placental trophoblasts, which, in turn, may stimulate endogenous CO production. The CO may inhibit placental apoptosis directly and, thus, reduces the release of placenta-derived toxic factors. The crosstalk between CO and HO in contributing the protection of smoker in preeclampsia is also likely. The initial study on the possible involvement of HO pathway in the pathogenesis of preeclampsia reported from human studies that have demonstrated lower HO gene expression in placental villous tissue during the first trimester among pregnant women destined to develop preeclampsia and decreased exhalation of CO among women with preeclampsia (Kreiser et al. 2004; Farina et al. 2008).

8.5 Effect of Caffeine on Placental Growth and Development and Fetus

Caffeine is actively known as methylxanthine derived naturally from the plants. Due to their wide popularity, these compounds are omnipresent and mixed with the beverages such as coffee, tea, cocoa/chocolate, and cola drinks, in many energy drinks as well as in the common psychostimulant. Caffeine is the most common xenobiotic consumed during pregnancy. Advising on restricted caffeine intake during pregnancy is universally accepted recommendation (Wadge 2009). The effects of caffeine on pregnancy outcome are controversial; yet, epidemiological studies have established a correlation between increased caffeine intake and risk of miscarriage and IUGR. Recent epidemiological study from Norwegian Mother and Child Cohort ($n = 59{,}123$), women with uncomplicated pregnancies with a live singleton babies were identified to examine the association between maternal caffeine intake at gestational intervals (week 17, 22, and 30) and risk for spontaneous preterm delivery (PTD), birth weight (BW), and small for gestational age (SGA) by questionnaire methods. Although, main caffeine was derived from coffee, tea and chocolate were preferred by women who wished low caffeine intake. Median pre-pregnancy caffeine intake, 126 mg/day, 44 mg/day (week 17), and 62 mg/day (week 30), was much lower than recommended maximum of 200 mg/day in the Nordic countries and 300 mg/day in USA according to the World Health Organization (WHO). Coffee caffeine, but not caffeine from other sources, was associated with prolonged gestation marginally. Neither total nor coffee caffeine was associated with spontaneous PTD risk. Caffeine intake was consistently associated with decreased BW, and increased odds of SGA in spite of ~50 % less intake than maximum recommendation suggests a further restriction on its intake during pregnancy (Sengpiel et al. 2013). Danish National Birth Cohort data ($n = 71{,}239$), 1996–2002 of nondiabetic women with singleton pregnancies, examined the relation between first trimester coffee and tea consumption and gestational diabetes mellitus (GDM) risk. Data suggest that moderate first trimester coffee and tea intakes were not associated with increased risk for GDM. Coffee or tea intake was reported in majority of the pregnant ($n = 57{,}882$) women (Hinkle et al. 2014).

In spite of the fact that caffeine intake sometime becomes unavoidable in the conception and preconception period due to unrecognized pregnancy, impact of caffeine on early trophoblast developmental function is not well reported. Caffeine is lipid soluble and diffuse freely across biological membranes, and their presence is detected in the preimplantation blastocyst and placenta. The primary cellular response upon caffeine intake is the repression of cAMP phosphodiesterase activity that results in the elevation of intracellular cAMP which can stimulate calcium release from intracellular stores and thereby inhibiting cellular proliferation and inducing cell cycle arrest and apoptosis. Caffeine can also inhibit tumor cell invasion (Gude et al. 2001). The membrane-permeable analog of cAMP mimicked the inhibitory effect of caffeine on EGF-stimulated trophoblast invasion of fibrin gels. Although the mechanism responsible for this is not known, evidence from other cell systems suggests that both caffeine and cAMP can inhibit the expression and/or activation of matrix metalloproteinase-2 and -9. Because both enzymes are expressed by trophoblasts and are active in regulating trophoblast invasion, their role in the regulation of the trophoblast invasion is speculated. Scanty data are available on the impact of caffeine on normal trophoblast development and its proliferation and uterine invasion involving the trophoblast cells. The effect of caffeine on trophoblast migration and motility at basal level and epidermal growth factor (EGF)-stimulated condition was studied using primary first-trimester extravillous trophoblasts (EVT) and the EVT-derived cell line SGHPL-4 to investigate the intracellular signaling pathways involved in this process. Caffeine at the level of 0.2 mM inhibited basal and EGF-stimulated trophoblast invasion and motility in a dose-dependent manner. The inhibition was largely independent of the generation of cAMP but was mediated via inactivation of PI3-kinase and mTORC2, resulting in a failure to activate Akt, a key intermediate in the intracellular pathway leading to trophoblast invasion and motility (Grant et al. 2012). Caffeine induced placental alteration of glucocorticoid precursor. Caffeine decreased placental 11beta-HSD2 activity dose dependently both at protein and mRNA level via adenosine A(2B) receptor-mediated induction, which suggests a functional link between exposure to caffeine and decreased intracellular levels of cAMP and reduced placental 11beta-HSD2 (Sharmin et al. 2012).

Caffeine is commonly consumed during pregnancy, crosses the placenta, with fetal serum concentrations similar to the mother's, but studies of birth outcome show conflicting findings. Considering the trend from many large cohorts recently, reductions in the maximum recommended intake of caffeine during pregnancy need additional studies (Greenwood et al. 2014). Till that time, minimum intake of caffeine would be a worthy advice one can suggest based on these studies.

8.6 Maternal Alcohol Intake on Hormone, Angiogenesis and Vascular System of the Placenta

Maternal alcohol ingestion during pregnancy produces adverse embryo/fetal development and gestation outcomes, collectively referred to as fetal alcohol spectrum disorders (FASDs). Consumption of alcohol rate is increasing worldwide that includes the women of reproductive age. Chronic alcohol intake during pregnancy is associated with FASD that includes a range of developmental defect, behavioral difficulty, and fetal growth restriction. Poor placentation that leads to FGR often manifested with lower fetus weight, increased placental apoptosis, reduced cellular growth, proliferation of the placenta, and superficial invasion of trophoblast into uterine spiral arteries. Available approved drug therapies for the treatment of FASD are limited.

Interestingly, alcohol utilization and metabolism between men and women are not similar as the first pass metabolism and gastric alcohol dehydrogenase activity are as low as 23 and 59 % in women which results in much higher blood alcohol level in women (Frezza et al. 1990). Several studies investigated alcohol effects on maternal hormones that are known to affect maternal resistance of blood vessels. Activity of hypothalamic pituitary axis (HPA) is linked with fetal alcohol syndrome (FAS). Studies have shown that maternal alcohol consumption is reported to increase maternal basal and stress HPA activity (Cudd et al. 2001). Alcohol consumption in women during pregnancy leads to lower levels of sex hormone-binding globulin throughout pregnancy and total testosterone concentrations at 16–20 weeks of gestation (term is roughly 285 days). Recent study have shown a nearly 25 % decrease in serum testosterone levels in women who drink a median of one glass a day (Stevens et al. 2005). Pregnant drinker who consumed higher amounts of alcohol (5- to 20-year history of alcohol abuse; consumption, 140–840 g/week during first half of gestation or beyond) showed lower ratio of the urinary combined prostacyclin metabolites (vasodilator) and thromboxane B2 (vasoconstrictor), compared with nonalcoholics, which indicates vasoconstriction and increased resistance in blood vessel. Since angiogenesis and vessel remodeling of the placental trophoblast are the major contributing mechanisms defining the utero-placental circulation during gestation, effects of alcohol on these processes employed animal models to mimic the situation. In the mother, chronic alcohol (37 % of caloric content) impairs the physiological conversion of the maternal uterine vasculature within the mesenteric triangle in rats, a necessary step for delivery of nutrients and gas to the fetus (Gundogan et al. 2008). In a chick extraembryonic model, moderate and heavy alcohol doses (30 or 50 %, 24 or 48 h) impaired vascular development which was attributed to deficits to angiogenic growth factors, associated receptors, and oxidative stress (Tufan and Satiroglu-Tufan 2003; Gu et al. 2001). Recent study analyzed genes of angiogenesis pathway following alcohol exposure. Among the genes analyzed for mRNA expression, 20 angiogenesis associated genes were downregulated and two upregulated (Ramadoss and Magness 2012). The downregulated genes primarily clustered as angiogenic growth factors/receptors (placental growth factor), adhesion molecules (angiopoietin-like-3, collagen-18A1, endoglin), proteases/matrix proteins/inhibitors (alanyl aminopeptidase, collagen-4A3, heparanase, plasminogen, platelet factor-4, plexin domain containing-1, tissue inhibitor of metalloproteinases-3), and miscellaneous growth factors such as leptin, platelet-derived growth factor-α, transforming growth factor-α, whereas plasminogen activator urokinase and transforming growth factor-β receptor-1 were upregulated. These data suggest that alcohol affects vascular adaptations during pregnancy by altering angiogenic indexes.

Ethanol effects in the embryo and gestation depend on doses and concentrations of ethanol, route of administration, and time and period of exposure (peri/pre). In spite of large differences in the trophoblast invasion system of mouse and human, the study on short-term periconceptional ethanol exposure on embryo growth, differentiation, implantation, and invasion was found interesting in mouse model to mimic those mothers exposed to this situation unknowingly. Recent study showed that perigestational ethanol (10 %) exposure 17 days prior to gestation and up to day 10 of gestation produced various abnormalities in embryo development that include reduced embryo viability, delayed embryo differentiation and growth, increased frequency of neural tube defects, and impaired growth promoting arachidonic acid metabolism in mouse (Cebral et al. 2007). In fact, low levels of ethanol reduced synthesis of PGE2 by blastocysts developed in vivo, so poor blastocyst differentiation may also probably be related to impaired levels of embryonic PGE. Prostaglandins are implicated in blastocyst expansion, participate in the hatching of blastocysts, and are involved in the initiation of implantation. Exposure of ethanol resulted in morphological abruption of blastocyst development with less number of nuclei in early blastocysts which indicates decreased embryonic size, low growth, and/or delayed or arrested mitotic cycle, suggesting a reduction of embryonic cell proliferation. The differentiation of embryo at early and postnatal phases was altered under the influence of alcohol. Possibility of spontaneous pregnancy loss was expected due to sharp decline in the number of embryo type 1 at the final phase of implantation. In ethanol-exposed females, a higher quantity of type 1 embryos was lost in earlier stages than in controls and did not survive beyond the end of the culture time, indicating an increased risk of early pregnancy loss during implantation (Pérez-Tito et al. 2013). However, the high number of type 2 embryos at all implantation stages suggests that this embryo population retained the highest survival rate up to the final phases of implantation. Moreover, the reduced quantity of type 3 embryos in ethanol-exposed females confirms retardation and/or deregulation of differentiation near implantation and suggests that periconceptional alcohol consumption may cause embryo retarded differentiation at early post-implantation phases. Overall, this study suggests that periconceptional alcohol exposure induces delayed embryo differentiation and stimulation of trophoblast expansion to increase their invasiveness during implantation. Impaired implantation is another key component of placentation that is likely to be altered in ethanol-exposed female mouse.

8.7 Effect of Maternal Alcohol Intake on Nutrient Transport and Insulin Signaling of the Placenta

Chronic exposure of high level of ethanol (20 %v/v) during pregnancy decreases fetal growth, pup development, and mortality in mouse. With high alcohol intake, placental development was significantly affected due to irregular trophoblast morphology, altered blood vessel formation in the labyrinth zone, and impaired amino acid transport to the fetus. Facial dysmorphia was typically observed in mouse even with moderate level of alcohol exposure (6 %v/v) during pregnancy. Alcohol and its metabolites such as acetaldehyde are readily diffused into the placenta depending on concentration and accumulate into the fetal blood at a similar concentration to those measured in maternal blood. In fact, the teratogenic effects of these metabolites are known to affect growth and development of the term fetus. The permissible limit of blood alcohol concentration should be less than 0.08 % by volume (~17.7 mM) for driving in the UK and USA. Literature on peak blood alcohol suggests that 40 mM causes intoxication in a normal population and equivalent to 3–4 standard drinks. The average peak blood acetaldehyde concentration is in the range 26–43 mM (Lui et al. 2014). The study when used alcohol (<40 mM) and its metabolites acetaldehyde at clinically relevant concentration to measure their effect on trophoblast function, concluded adverse impact on two key aspects of trophoblast development, i.e., proliferation and transport function. In rat, ethanol exposure reduces the number of invasive trophoblast giant cells and induces cell death in the spongiotrophoblast layer. Defective invasion of extravillous cytotrophoblasts with reduced arterial remodeling was predicted. In addition to trophoblast growth function, alcohol possibly modulates expression of trophoblast transporters. The placental system β amino acid transporter is key supplier to the fetus with taurine, an amino acid with antioxidant properties that may have neuroprotective effect during fetal development (Desforges et al. 2013b). Reduced placental taurine transporter (TauT) activity in pregnancies complicated by preeclampsia and maternal obesity was reported recently (Desforges et al. 2013a). Interestingly, taurine transport function was disrupted at low blood alcohol (0.05 %). Maternal taurine deprived animal model has lower birth weight, decreased brain weight compared to normal. It is not clear how alcohol intake alter transport function of this amino acid selectively. In vitro data at 40 mM alcohol suggest inhibition of taurine transporter function that may contribute towards neurological, behavioral, and cognitive differences of the child born to mothers exposed to this condition (Lui et al. 2014; Desforges et al. 2013b).

The mechanism of alcohol action on placentation is multifactorial at cellular and structural levels. The toxic effects of ethanol and its metabolites can be mediated either directly or indirectly since ethanol from the maternal blood readily diffuse to the fetus and placenta. Ethanol-induced reductions within the mass of the labyrinthine layer, a site where maximum nutrient exchange between the fetus and mother occurs, could impair nutrient delivery to the fetus, and thereby result in increased fetal demise as well as IUGR. This has been shown that folic acid transport to the human fetus is decreased in pregnancies with chronic alcohol exposure (Hutson et al. 2012). Serum folate was measured in maternal blood and umbilical cord blood at the time of delivery in pregnancies with chronic and heavy alcohol exposure ($n = 23$) and in nondrinking controls ($n = 24$). In the alcohol-exposed pairs, the fetal:maternal serum folate ratio was ≤ 1.0 in more than 50 % of the subjects ($n = 14$) which indicates altered transport of folate to the fetus upon heavy alcohol intake. Authors presume that altered folate concentrations within the placenta and in the fetus may in part contribute to the deficits observed in the fetal alcohol spectrum disorders. A second major placental abnormality associated with chronic gestational exposure to ethanol was reduced transformation of maternal uterine blood vessels within the mesometrial triangle, i.e., the deep placental bed (Gundogan et al. 2008). Insulin growth factor (IGF)-mediated signaling is also target of maternal alcohol toxicity. IGF-1 and IGF-2 activations, which are required for the growth of the fetus and survival signaling mechanisms, are impaired by ethanol. IGF-2 transmits its vital signals by activating PI3 kinase-Akt and Erk-MAPK via its own receptor or by binding to the insulin or IGF-1 receptors. The elevated level of IGF-2 and IGF-2 receptor expression in ethanol-exposed placentas was a compensatory response to impairments in IGF-1-mediated signaling and thereby compromise the growth and survival pathways of the fetus. The other possible mediator of ethanol action on IGF signaling could be the level of aspartyl-(asparaginyl) β-hydroxylase (AAH) gene that is regulated by IGF signaling pathway (Gundogan et al. 2007). AAH has a critical role in cell motility and invasion which are required for successful placentation. Extravillous trophoblastic cells that mediate the placentation and vascular transformation were immunoreactive for AAH as well as cytokeratin 7, which is a trophoblastic cell marker. AAH protein levels were reduced in ethanol-exposed placentas. Since AAH has an important role in motility and invasion, ethanol-mediated inhibition of the AAH gene most likely represents an important mechanism of ethanol-impaired placentation.

8.8 Summary

The protection, if any, offered by HO/CO/AM to the women with a preeclampsia or not, the final call to adopt this approach of cigarette smoking must consider the adverse outcomes of the pregnancy and fetus growth restriction (Fig. 8.1). Whether protective effect of smoking could be exploited as a therapy in managing the predictive cases of preeclampsia (based on history) may depend on evidences based on first trimester placenta instead of term placenta, with a clear end user benefit for both mother and fetus which outweighs any risk before a treatment is used in clinical practice. The case of smoking in the protection of preeclampsia seems to be restricted at maternal level and no studies yet correlate fetal outcome and birth weight. Considering the trend from many large cohorts recently, reductions in the maximum recommended intake of caffeine during pregnancy need additional studies (Greenwood et al. 2014). Till that time, minimum intake of

Fig. 8.1 Hypothetical protective effect of cigarette smoke extract on risk factor of preeclampsia. *HO* hemoxygenase, *CO* carbon monoxide, *AM* adrenomedullin

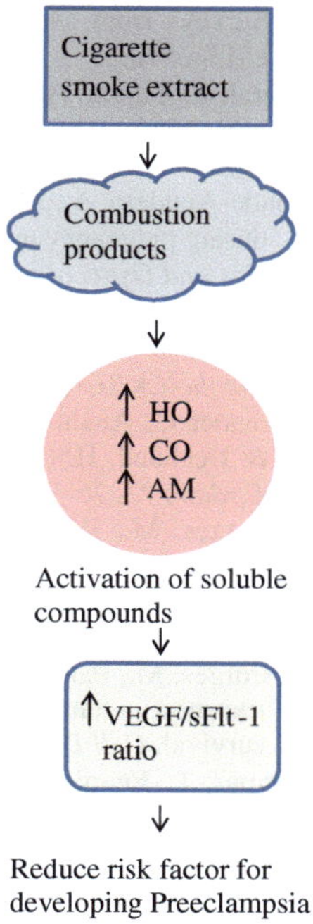

caffeine would be a worthy advice one can suggest based on these studies. Understanding the effect of smoking and alcohol-related toxicity on the uptake of the nutrients which are important for placental growth and development such as amino acids, glucose, DHA, and folic acid would be important to ascertain how these environmental factors affect the process of placentation in utero.

References

Albuquerque, C. A., Smith, K. R., Johnson, C., Chao, R., & Harding, R. (2004). Influence of maternal tobacco smoking during pregnancy on uterine, umbilical and fetal cerebral artery blood flows. *Early Human Development, 80*, 31–42.

Bainbridge, S. A., Sidle, E. H., & Smith, G. N. (2005). Direct placental effects of cigarette smoke protect women from pre-eclampsia: The specific roles of carbon monoxide and antioxidant systems in the placenta. *Medical Hypotheses, 64*, 17–27.

Belgore, F. M., Lip, G. Y., & Blann, A. D. (2000). Vascular endothelial growth factor and its receptor, Flt-1, in smokers and non-smokers. *British Journal of Biomedical Science, 57*, 207–213.

Bruchova, H., Vasikova, A., Merkerova, M., Milcova, A., Topinka, J., Balascak, I., Pastorkova, A., Sram, R. J., & Brdicka, R. (2010). Effect of maternal tobacco smoke exposure on the placental transcriptome. *Placenta, 31*, 186–191.

Caleyachetty, R., Tait, C. A., Kengne, A. P., Corvalan, C., Uauy, R., & Echouffo-Tcheugui, J. B. (2014). Tobacco use in pregnant women: Analysis of data from Demographic and Health

Surveys from 54 low-income and middle-income countries. *The Lancet Global Health, 2*, e513–e520.
Cebral, E., Faletti, A., Jawerbaum, A., & Paz, D. (2007). Periconceptional alcohol consumption-induced changes in embryonic prostaglandin E levels in mouse organogenesis: Modulation by nitric oxide. *Prostaglandins, Leukotrienes, and Essential Fatty Acids, 76*, 141–151.
Conde-Agudelo, A., Althabe, F., Belizan, J. M., & Kafury-Goeta, A. C. (1999). Cigarette smoking during pregnancy and risk of preeclampsia: A systematic review. *American Journal of Obstetrics and Gynecology, 181*, 1026–1035.
Cudd, T. A., Chen, W. J., & West, J. R. (2001). Fetal and maternal sheep hypothalamus pituitary adrenal axis responses to chronic binge ethanol exposure during the third trimester equivalent. *Alcoholism: Clinical and Experimental Research, 25*, 1065–1071.
Dechanet, C., Anahory, T., Daude, J. C. M., Quantin, X., Reyftmann, L., Hamamah, S., Hedon, B., & Dechaud, H. (2011). Effects of cigarette smoking on reproduction. *Human Reproduction Update, 17*, 76–95.
Desforges, M., Ditchfield, A., Hirst, C. R., Pegorie, C., Martyn-Smith, K., Sibley, C. P., & Greenwood, S. L. (2013a). Reduced placental taurine transporter (TauT) activity in pregnancies complicated by pre-eclampsia and maternal obesity. *Advances in Experimental Medicine and Biology, 776*, 81–91.
Desforges, M., Parsons, L., Westwood, M., Sibley, C. P., & Greenwood, S. L. (2013b). Taurine transport in human placental trophoblast is important for regulation of cell differentiation and survival. *Cell Death and Disease, 4*, e559.
Detmar, J., Rennie, M. Y., Whiteley, K. J., Qu, D., Taniuchi, Y., Shang, X., Casper, R. F., Adamson, S. L., Sled, J. G., & Jurisicova, A. (2008). Fetal growth restriction triggered by polycyclic aromatic hydrocarbons is associated with altered placental vasculature and AhR-dependent changes in cell death. *American Journal of Physiology, Endocrinology and Metabolism, 295*, E519–E530.
Dikensoy, E., Balat, O., Pence, S., Balat, A., Cekmen, M., & Yurekli, M. (2009). The changes of plasma malondialdehyde, nitric oxide, and adrenomedullin levels in patients with preeclampsia. *Hypertension in Pregnancy, 28*, 383–389.
Farina, A., Sekizawa, A., de Sanctis, P., Purwosunu, Y., Okai, T., Cha, D. H., Kang, J. H., Vicenzi, C., Tempesta, A., Wibowo, N., Valvassori, L., & Rizzo, N. (2008). Gene expression in chorionic villous samples at 11 weeks' gestation from women destined to develop preeclampsia. *Prenatal Diagnosis, 28*, 956–961.
Frezza, M., di Padova, C., Pozzato, G., Terpin, M., Baraona, E., & Lieber, C. S. (1990). High blood alcohol levels in women. The role of decreased gastric alcohol dehydrogenase activity and first-pass metabolism. *The New England Journal of Medicine, 322*, 95–99.
Genbacev, O., Bass, K. E., Joslin, R. J., & Fisher, S. J. (1995). Maternal smoking inhibits early human cytotrophoblast differentiation. *Reproductive Toxicology (Elmsford, NY), 9*, 245–255.
Gladen, B. C., Zadorozhnaja, T. D., Chislovska, N., Hryhorczuk, D. O., Kennicutt, M. C., II, & Little, R. E. (2000). Polycyclic aromatic hydrocarbons in placenta. *Human and Experimental Toxicology, 19*, 597–603.
Grant, I., Cartwright, J. E., Lumicisi, B., Wallace, A. E., & Whitley, G. S. (2012). Caffeine inhibits EGF-stimulated trophoblast cell motility through the inhibition of mTORC2 and Akt. *Endocrinology, 153*, 4502–4510.
Gratton, R. J., Gluszynski, M., Mazzuca, D. M., Nygard, K., & Han, V. K. (2003). Adrenomedullin messenger ribonucleic acid expression in the placentae of normal and preeclamptic pregnancies. *Journal of Clinical Endocrinology and Metabolism, 88*, 6048–6055.
Greenwood, D. C., Thatcher, N. J., Ye, J., Garrard, L., Keogh, G., King, L. G., & Cade, J. E. (2014). Caffeine intake during pregnancy and adverse birth outcomes: A systematic review and dose-response meta-analysis. *European Journal of Epidemiology, 29*, 725–734.
Gu, J. W., Elam, J., Sartin, A., Li, W., Roach, R., & Adair, T. H. (2001). Moderate levels of ethanol induce expression of vascular endothelial growth factor and stimulate angiogenesis. *American Journal of Physiology. Regulatory, Integrative and Comparative Physiology, 281*, R365–R372.

Gude, R. P., Menon, L. G., & Rao, S. G. (2001). Effect of Caffeine, a xanthine derivative, in the inhibition of experimental lung metastasis induced by B16F10 melanoma cells. *Journal of Experimental & Clinical Cancer Research, 20*, 287–292.

Gundogan, F., Elwood, G., Greco, D., Rubin, L. P., Pinar, H., Carlson, R. I., Wands, J. R., & de la Monte, S. M. (2007). Role of aspartyl-(asparaginyl) beta-hydroxylase in placental implantation: Relevance to early pregnancy loss. *Human Pathology, 38*, 50–59.

Gundogan, F., Elwood, G., Longato, L., Tong, M., Feijoo, A., Carlson, R. I., Wands, J. R., & de la Monte, S. M. (2008). Impaired placentation in fetal alcohol syndrome. *Placenta, 29*, 148–157.

Hakkola, J., Pasanen, M., Pelkonen, O., Hukkanen, J., Evisalmi, S., Anttila, S., Rane, A., Mantyla, M., Purkunen, R., Saarikoski, S., Tooming, M., & Raunio, H. (1997). Expression of CYP1B1 in human adult and fetal tissues and differential inducibility of CYP1B1 and CYP1A1 by Ah receptor ligands in human placenta and cultured cells. *Carcinogenesis, 18*, 391–397.

Hammoud, A. O., Bujold, E., Sorokin, Y., Schild, C., Krapp, M., & Baumann, P. (2005). Smoking in pregnancy revisited: Findings from a large population-based study. *American Journal of Obstetrics and Gynecology, 192*, 1856–1862. discussion 1862–1863.

Hinkle, S., Laughon, S., Catov, J., Olsen, J., & Bech, B. (2014). First trimester coffee and tea intake and risk of gestational diabetes mellitus: A study within a national birth cohort. *BJOG, 122*(3), 420–428.

Hutson, J. R., Stade, B., Lehotay, D. C., Collier, C. P., & Kapur, B. M. (2012). Folic acid transport to the human fetus is decreased in pregnancies with chronic alcohol exposure. *PLoS One, 7*, e38057.

Jauniaux, E., & Burton, G. J. (2007). Morphological and biological effects of maternal exposure to tobacco smoke on the feto-placental unit. *Early Human Development, 83*, 699–706.

Kamarainen, M., Soini, T., Wathen, K. A., Leinonen, E., Stenman, U. H., & Vuorela, P. (2009). Smoking and sVEGFR-1: Circulating maternal concentrations and placental expression. *Molecular and Cellular Endocrinology, 299*, 261–265.

Karumanchi, S. A., & Levine, R. J. (2010). How does smoking reduce the risk of preeclampsia? *Hypertension, 55*, 1100–1101.

Kimya, Y., Cengiz, C., Ozan, H., & Kolsal, N. (1998). Acute effects of maternal smoking on the uterine and umbilical artery blood velocity waveforms. *Journal of Maternal-Fetal Investigation, 8*, 79–81.

Kitamura, M., & Kasai, A. (2007). Cigarette smoke as a trigger for the dioxin receptor-mediated signaling pathway. *Cancer Letters, 252*, 184–194.

Kraus, D. M., Feng, L., Heine, R. P., Brown, H. L., Caron, K. M., Murtha, A. P., & Grotegut, C. A. (2014). Cigarette smoke-induced placental adrenomedullin expression and trophoblast cell invasion. *Reproductive Sciences (Thousand Oaks, CA), 21*, 63–71.

Kreiser, D., Baum, M., Seidman, D. S., Fanaroff, A., Shah, D., Hendler, I., Stevenson, D. K., Schiff, E., & Druzin, M. L. (2004). End tidal carbon monoxide levels are lower in women with gestational hypertension and pre-eclampsia. *Journal of Perinatology, 24*, 213–217.

Lain, K. Y., Powers, R. W., Krohn, M. A., Ness, R. B., Crombleholme, W. R., & Roberts, J. M. (1999). Urinary cotinine concentration confirms the reduced risk of preeclampsia with tobacco exposure. *American Journal of Obstetrics and Gynecology, 181*, 1192–1196.

Larsen, L. G., Clausen, H. V., & Jonsson, L. (2002). Stereologic examination of placentas from mothers who smoke during pregnancy. *American Journal of Obstetrics and Gynecology, 186*, 531–537.

Lenhart, P. M., & Caron, K. M. (2012). Adrenomedullin and pregnancy: Perspectives from animal models to humans. *Trends in Endocrinology and Metabolism, 23*, 524–532.

Li, M., Li, M., Schwerbrock, N. M., Lenhart, P. M., Fritz-Six, K. L., Kadmiel, M., Christine, K. S., Kraus, D. M., Espenschied, S. T., Willcockson, H. H., Mack, C. P., & Caron, K. M. (2013). Fetal-derived adrenomedullin mediates the innate immune milieu of the placenta. *Journal of Clinical Investigation, 123*, 2408–2420.

Lui, S., Jones, R. L., Robinson, N. J., Greenwood, S. L., Aplin, J. D., & Tower, C. L. (2014). Detrimental effects of ethanol and its metabolite acetaldehyde, on first trimester human placental cell turnover and function. *PLoS One, 9*, e87328.

Mehendale, R., Hibbard, J., Fazleabas, A., & Leach, R. (2007). Placental angiogenesis markers sFlt-1 and PlGF: Response to cigarette smoke. *American Journal of Obstetrics and Gynecology, 197*, 363.e1–363.e5.

Morrow, R. J., Ritchie, J. W., & Bull, S. B. (1988). Maternal cigarette smoking: the effects on umbilical and uterine blood flow velocity. *American Journal of Obstetrics and Gynecology, 159*, 1069–1071.

Myllynen, P., Immonen, E., Kummu, M., & Vahakangas, K. (2009). Developmental expression of drug metabolizing enzymes and transporter proteins in human placenta and fetal tissues. *Expert Opinion on Drug Metabolism and Toxicology, 5*, 1483–1499.

Pérez-Tito, L., Bevilacqua, E., & Cebral, E. (2013). Peri-implantational in vivo and in vitro embryo-trophoblast development after perigestational alcohol exposure in the CD-1 mouse. *Drug and Chemical Toxicology, 37*(2), 184–197.

Powers, R. W., Roberts, J. M., Cooper, K. M., Gallaher, M. J., Frank, M. P., Harger, G. F., & Ness, R. B. (2005). Maternal serum soluble fms-like tyrosine kinase 1 concentrations are not increased in early pregnancy and decrease more slowly postpartum in women who develop preeclampsia. *American Journal of Obstetrics and Gynecology, 193*, 185–191.

Ramadoss, J., & Magness, R. R. (2012). Multiplexed digital quantification of binge-like alcohol-mediated alterations in maternal uterine angiogenic mRNA transcriptome. *Physiological Genomics, 44*, 622–628.

Rennie, M. Y., Detmar, J., Whiteley, K. J., Yang, J., Jurisicova, A., Adamson, S. L., & Sled, J. G. (2011). Vessel tortuousity and reduced vascularization in the fetoplacental arterial tree after maternal exposure to polycyclic aromatic hydrocarbons. *American Journal of Physiology. Heart and Circulatory Physiology, 300*, H675–H684.

Schmidt-Lucke, C., Belgore, F., Reinhold, D., Ansorge, S., Klein, H. U., Schmidt-Lucke, J. A., & Lip, G. Y. (2005). Soluble vascular endothelial growth factor, soluble VEGF receptor Flt-1 and endothelial function in healthy smokers. *International Journal of Cardiology, 100*, 207–212.

Sengpiel, V., Elind, E., Bacelis, J., Nilsson, S., Grove, J., Myhre, R., Haugen, M., Meltzer, H. M., Alexander, J., Jacobsson, B., & Brantsaeter, A. L. (2013). Maternal caffeine intake during pregnancy is associated with birth weight but not with gestational length: Results from a large prospective observational cohort study. *BMC Medicine, 11*, 42.

Sharmin, S., Guan, H., Williams, A. S., & Yang, K. (2012). Caffeine reduces 11beta-hydroxysteroid dehydrogenase type 2 expression in human trophoblast cells through the adenosine A(2B) receptor. *PLoS One, 7*, e38082.

Shinjo, A., Ventura, W., Koide, K., Hori, K., Yotsumoto, J., Matsuoka, R., Ichizuka, K., & Sekizawa, A. (2014). Maternal smoking and placental expression of a panel of genes related to angiogenesis and oxidative stress in early pregnancy. *Fetal Diagnosis and Therapy, 35*(4), 289–95.

Stejskalova, L., Vecerova, L., Peréz, L. M., Vrzal, R., Dvorak, Z., Nachtigal, P., & Pavek, P. (2011). Aryl hydrocarbon receptor and aryl hydrocarbon nuclear translocator expression in human and rat placentas and transcription activity in human trophoblast cultures. *Toxicological Sciences, 123*, 26–36.

Stevens, R. G., Cohen, R. D., Terry, M. B., Lasley, B. L., Siiteri, P., & Cohn, B. A. (2005). Alcohol consumption and serum hormone levels during pregnancy. *Alcohol, 36*, 47–53.

Tufan, A. C., & Satiroglu-Tufan, N. L. (2003). The effect of ethanol exposure on extraembryonic vascular development in the chick area vasculosa. *Cells, Tissues, Organs, 175*, 84–97.

Wadge, A. (2009). Caffeine in pregnancy. Food Standard Agency's advice on caffeine. *BMJ, 338*, b299.

Wang, K., Zhou, Q., He, Q., Tong, G., Zhao, Z., & Duan, T. (2011). The possible role of AhR in the protective effects of cigarette smoke on preeclampsia. *Medical Hypotheses, 77*, 872–874.

Wikstrom, A. K., Stephansson, O., & Cnattingius, S. (2010). Tobacco use during pregnancy and preeclampsia risk: Effects of cigarette smoking and snuff. *Hypertension, 55*, 1254–1259.

Zhang, J., Klebanoff, M. A., Levine, R. J., Puri, M., & Moyer, P. (1999). The puzzling association between smoking and hypertension during pregnancy. *American Journal of Obstetrics and Gynecology, 181*, 1407–1413.

Chapter 9
Regulation of Placentation by Environmental Factors

Contents

9.1 Introduction

Changes in the maternal environment can impair fetal growth and development, which may result in increased susceptibility to diseases in postnatal life. The placental nutrient sensing capacity may integrate these perturbations with information from intrinsic nutrient sensing signaling pathways to regulate secretion of hormones and placental nutrients and oxygen transfer. Placental responses to perturbations in the maternal compartment are complex and poorly understood, highlighting an urgent need for further well-designed and mechanistic research in this area. Intervention strategies to alleviate pregnancy complications and prevent fetal programing of adult disease are likely to be the most effective if placental function is targeted. The effects of environment exposure pollutants on placentation is warranted since pregnant mothers are often exposed to these factors during preconception to gestational period. Impaired placentation that ultimately influences fetus outcome is the center of many development disorders and contributes as an independent risk factor for the adult chronic diseases. The interaction of genes and the environment that predict the fetus outcome was evidenced where birth weight was reduced in babies derived from the women smoked during pregnancy irrespective of their genotypes. While susceptible genotypes in addition to smoking contributed further reduction in the birth weight

A.K. Duttaroy, S. Basak, *Early Nutrition and Lifestyle Factors*,
DOI 10.1007/978-3-319-38804-5_9

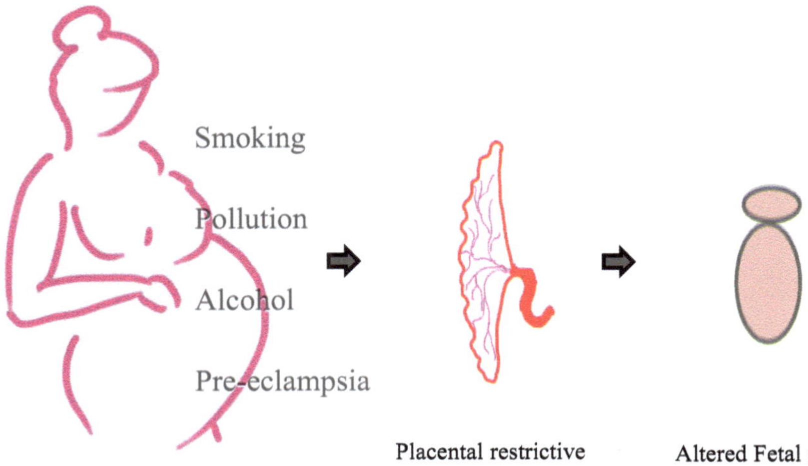

Fig. 9.1 Effect of maternal exposure of environmental factors on feto-placental growth and development

explained why we need to measure the quantum of changes in genotypes that comes from the lifestyle habit, i.e., smoking or alcohol, etc. during pregnancy (Fig. 9.1). Placental gene expression affects its invasive and functional capacities and thereby the growth of the feto-placental unit. Placental secreted factors that influence the maternal adaptation to pregnancy and, hence, the health of both mother and baby are influenced by the stress factors. Many studies raised the issue that prenatal exposure to environmental pollutants that include air pollutants, food contaminants, and chemicals present in consumer products may contribute additional insult to the early developmental process in utero. The biomarker measurements of such conditions by mimicking their effect on placental growth and development and gene expression may provide a new insight into the association between early exposure of these modifiable environmental factors and fetus outcome. This chapter will review the subtle effect induced by environmental risk factors exposed to mother during pregnancy which can lead to functional insufficiency in the placenta due to alteration in the gene expression of the functional pathways associated with placentation. Impact of environmental toxins on placental blood flow, vasculature, and angiogenesis of the placenta are also discussed.

9.2 Effects of Environmental Pollutants on Angiogenic and Vascular Development of the Placenta

Epidemiologic studies suggest that prenatal exposure to air pollutants, several food contaminants, and chemicals present in consumer products are associated with nongenetically transmitted adverse health effects, which manifest after birth

(Schoeters et al. 2011). Changes in cognitive behavior, sexual development, the prevalence of asthma and allergy, and growth curves have been shown to be associated with pollutant exposure at early life stages. Biomarkers of exposure can be measured in mothers before conception, during pregnancy, or after birth. Different biological tissues—such as peripheral or cord blood samples, hair samples, meconium, and urine—provide specific information that reflects the actual exposure dose during pregnancy or at birth. Biomarkers of effect may include changes in hormone concentrations, oxidative stress variables, changes in gene expression levels, and epigenetic changes.

The small cohort of normotensive pregnancies ($n = 22$) involving mothers and fetus investigated the effect of the polychlorinated biphenyl (PCB) exposure on the factors associated with placental function, including placental angiogenesis factors such as PlGF and sFlt-1 levels, and estimated total syncytiotrophoblast volume (Tsuji et al. 2013). Detected PCB values were more in maternal than cord blood, on a lipid or wet-weight basis, with maternal blood > cord blood. PCBs with a greater number of attached chlorine atoms had a lower rate of crossing the placental barrier from maternal circulation to fetal circulation. The PlGF/ST ratio was determined for individual placentas and compared to PCB exposure. A significant relationship between total PCB concentration and PlGF/syncytiotrophoblast was found for both maternal and cord blood. PCB exposure did not influence changes in the birth weight of the babies in spite of significant alteration in the placental syncytiotrophoblast volume and placental expression of growth factors. Therefore, placental plasticity against the environmental challenges and adaptive response to the challenges are important to maintain placental function for the successful fetus outcome. However, long-term effect of this alteration on genotype of the fetus would be required to estimate its trans-generational effect on fetus outcome. In utero exposure to an environmental toxicant, TCDD (2,3,7,8-tetrachlorodibenzo-*p*-dioxin), induces the intrauterine fetal death in many species via the activation of aryl hydrocarbon receptor (AhR). The effects of 2-(1′H-indole-3′-carbonyl)-thiazole-4-carboxylic acid methyl ester (ITE) and TCDD on the vascular remodeling of rat placentas detected (Wu et al. 2014) significant differences in the expression of angiogenic growth factors between ITE and TCDD-exposed groups, especially angiopoietin-2 (Ang-2), endoglin, interferon-gamma (IFN-gamma), and placental growth factor (PlGF). These results suggest that ITE and TCDD differentially regulate the vascular remodeling of rat placentas, as well as the expression of angiogenic factors and their receptors, which in turn may alter the blood flow in the late gestation and partially resulted in intrauterine fetal death.

Aryl hydrocarbon receptor (AhR) activated by TCDD triggers its downstream signaling pathway to exert adverse effects on vasculature development, angiogenesis, or vascular remodeling, in a variety of animals. The placenta, a mammalian organ rich in vasculature, consists of endothelial and trophoblast cells of fetal origin, which proliferate and differentiate under hypoxic condition in the uterine horn. A crosstalk between the hypoxia-inducible factor (HIF)-mediated pathway and AhR-mediated pathway is considered to play an important role in this physiological process. The toxicity of TCDD is mediated by the binding of TCDD to the aryl hydrocarbon receptor (AhR), which is then activated (Mimura and Fujii-Kuriyama 2003). The activated ligand-bound AhR translocates to the nucleus from the cytoplasm and dimerizes with the aryl hydrocarbon receptor nuclear translocator (ARNT), followed by the binding of this AhR/ARNT heterodimer to the xenobiotic response element (XRE; also known as the dioxin response element, DRE) in the promoter region of various genes. In functional XRE elements, the AhR/ARNT heterodimer modulates the expression of CYP1A1 and other genes for biological and toxicological responses. Vascular development is basically stimulated under a hypoxic condition, which is dependent on transcription factors known as the hypoxia-inducible factors (HIFs). Under normal oxygen tension, HIF-1α is posttranslationally modified and subsequently degraded through the proteasome. However, under hypoxic conditions, HIF-1α can escape from degradation, and the accumulated HIF-1α binds to the oxygen-insensitive molecule known as the ARNT, also called HIF-1β. The HIF-1α/ARNT heterodimer subsequently binds to the hypoxia response element (HRE) in the promoter region of genes involved in the adaptation to hypoxia. Thus, the HIF-1α acts as a master regulator to activate the transcription of many hypoxia-response genes, including the VEGF/VEGFR or Ang/Tie2 system, and regulates the expressions of VEGF, Flt1, and Ang2. HIF-1α plays its intrinsic role in hypoxia signaling and presumably modulates dioxin toxicities because of its ability to heterodimerize with ARNT. In other words, ARNT is a common transcription factor that shares its role with AhR and HIF-1α to modulate XRE- and HRE-dependent pathways, respectively. Thus, it has been speculated that the AhR/ARNT and HIF-1α/ARNT pathways affect each other by competing for the limited quantities of ARNT molecules (Okada et al. 2009).

AhR plays an important role as a mediator of an adaptive response to xenobiotics, as well as in normal physiology and embryonic development. Several exogenous AhR ligands, such as PAHs, polychlorinated biphenyls, and halogenated dioxins, can be found in the constituents of numerous commercial products, including insulators and flame retardants, or as products of combustion processes, including chimney soot, charbroiled foods, and cigarette smoke, or as the product of waste incineration. Exposure to these compounds subsequently affects cellular growth and differentiation, homeostasis, level of growth factors, reproductive function, and hormonal regulation. Importantly, elevated CYP1A1 activity through activated AhR in placentas of women smokers has been associated with pregnancy complications, such as premature birth, IUGR, structural abnormalities, fetal death or placenta abruption, risk of low birth weight, and low head circumference (Stejskalova and Pavek 2011). AhR protein level was found indifferent between the placenta derived from PE and healthy mothers. The AhR staining was evidenced in endothelium of large blood vessels in villi and endothelium of umbilical cord arteries and veins. In fetal tissues, the AhR immunoreactivity was localized in lung, kidney, esophagus, pancreas, liver, testicle, thymus gland, retina, and choroid, mainly in epithelial cells, whereas it was absent in heart, brain, sclera, and thoracic aorta (Jiang et al. 2010).

9.3 Placenta as an Environmental Biomarker for in Utero Exposure

Persistent organic pollutants (POPs) are ubiquitous chemicals that can accumulate in the human body because of their lipid solubility, slow persistence, and resistance to metabolism. Common POPs include organochlorine pesticides (OCPs), polychlorinated biphenyls (PCBs), polybrominated diphenyl ethers (PBDEs), and polycyclic aromatic hydrocarbons (PAHs), among others. Exposure to POPs can be associated with a wide range of adverse health effects that includes the disruption of normal embryonic development. Majority of these compounds act as endocrine disruptors, but they also have the ability to induce cellular oxidative stress. Elevated levels of POPs in the placenta was associated with increased risks of neural tube defect (NTD) as evidenced by increased PAHs, OCPs, PCBs, and PBDEs in the placentas ($n = 130$) in a rural area of China where NTD prevalence rate is as high as 13.9 per 1000 births (Ren et al. 2011). PAH concentrations were associated with a 4.52-fold increased risk for any NTDs and 5.84- and 3.71-fold increased risks for anencephaly and spina bifida, respectively. Placental levels of pollutants can be used as biomarkers of early in utero exposures even before the embryo proper is established. The effect of background exposure of the PCBs and dioxins on birth size was considered to the cases where in children born of mothers exposed to accidental high levels of PCBs and related compounds and in children born of mothers who consumed PCB-contaminated fish. Birth weight and growth rate were monitored in children ($n = 207$) at different points of time postnatally up to 42 months, whose mothers were exposed to PCBs (Patandin et al. 1998). The effect of in utero exposure to PCBs on birth size, assessed by cord and maternal plasma PCB levels, suggested that prenatal PCB exposure did not affect growth rate from 3 to 42 months of age. However, in utero exposure to environmental levels of PCBs is negatively associated with birth weight and postnatal growth until 3 months of age. This suggests that in utero alteration in the growth-promoting factors, and this possibly, involves the role of placenta in determining the feto-placental growth and outcome. The effects of ambient air pollution on hypertensive disorders in pregnancy and associations with obesity were conducted in a retrospective, case–control study ($n = 298$) in Hispanic women. Trimester-specific carbon monoxide (CO), nitrogen dioxide (NO_2), ozone (O_3), and particulate matter (PM) with aerodynamic diameter 10 μm and 2.5 μm (PM 10, PM 2.5) exposures were estimated based on 24-h exposure level at residential address. The first trimester exposure to PM 2.5 and carbon monoxide is associated with increased odds of hypertensive disorder of pregnancy independent of obesity (Mobasher et al. 2013).

9.4 Heavy Metal Contaminations and Feto-Placental Outcome

The toxicity of heavy metals on human health has been increasingly realized. Heavy metal exposure during prenatal life can have permanent damaging effects on human development and health. Heavy metal exposures during preconception can affect the feto-placental development by altering placentation processes. Preventive measures to eliminate or minimize the unnecessary risk of exposure to heavy metals or other pollutants during pregnancy should be initiated once these factors are identified. Heavy metals once in maternal blood can pass through the placenta and are directly deposited in growing fetal tissues. The damaging effects on the fetus depend on the amount of the teratogenic agents, the duration of the ingestion, the stage of placental development, the lipid solubility and molecular weight of the substance and, crucially, the stage of fetal life in which the substance was ingested. Teratogenic agents are most harmful during the 3rd to 12th weeks of fetal life when organ differentiation is occurring but, nonetheless, at any stage of the pregnancy, harmful effects may be exerted.

Cadmium (Cd) is a common environmental pollutant that mothers can potentially be exposed through tobacco smoke, diet, and occupation. Once absorbed, Cd has a long half-life in the body especially in the kidney and bones. Cd is known as a potent endocrine disruptor and, thus, may affect reproductive performance during pregnancy. Women are more susceptible to Cd toxicity due to increased intestinal uptake of Cd under low iron store condition. Women have higher blood cadmium levels than men from the same exposure possibly due to lower iron status. The placental cadmium level varies from 300 nM in potentially unexposed women to ~1 μM in potentially medium exposed women to ~2.7 μM in heavily exposed women. Indeed, mothers with placental concentrations (and higher) were reported to give birth to growth-retarded babies, and during pregnancy much of this cadmium accumulates in the placenta. Maternal exposure to Cd was associated with birth weight of the female offspring. A prospective cohort in a population-based nutritional supplementation trial in pregnancy was conducted in rural Bangladesh (Kippler et al. 2012). Women who had a singleton birth with measurements of size at birth and had donated a urine sample in early pregnancy for Cd analyses ($n = 1616$) were selected. Maternal urinary Cd (median, 0.63 μg/L) was significantly negatively associated with birth weight and head circumference. However, associations appeared to be limited to girls, with little evidence of effects in boys. More recent data have shown that the expression and activity of 11 beta-hydroxysteroid dehydrogenase type 2 (11beta-HSD2), which is causally linked to FGR, are decreased in cultured human trophoblast cells exposed to cadmium. Low level of cadmium exposure (0.5–1 μM $CdCl_2$) on placental function was investigated in vitro in the first trimester trophoblast cells by measuring migration activity of the immortalized human trophoblast cell line, HTR-8/SVneo. Cadmium at this level did not affect viability and basal migratory capacity of the HTR8/SVneo cells but suppressed cell migration induced by IGF-2 or PGE2. In addition to this, filamentous actin disorganization of the trophoblast cells was reported in the presence of cadmium. Actin cytoskeletal reorganization is a key phenomenon in the regulation of most of the above steps of cell migration. The study suggests that low level of Cd2+ did not affect trophoblast cell survival but interfered with ligand-induced trophoblast cell migration by affecting actin cytoskeletal organization (Alvarez and Chakraborty 2011). The placentation includes cell migration which is a multistep process involving many signaling molecules coordinating adhesion, projection of contraction, and cell detachment of the trophoblast cells. Improper function of one or more of the contributing components causes a reduced ability of these cells to establish an appropriate placental connection in situ, contributing to the increased incidence of fetal growth restriction in women exposed to cadmium.

9.5 Phthalate Contaminations and Its Effects on Feto-Placental Development

Phthalate is a synthetic compound used in plastics, personal care products, and building materials. People expose to phthalate in every day's life. However, this plasticizer excretes in urine 12–18 h after exposure. The phthalate metabolites when exposed to rats in utero, altered steroidogenesis and produced a phenotype with reduced testosterone synthesis in fetal Leydig cells (Barlow et al. 2003). In spite of the findings on phthalate-induced perturbation of steroidogenesis in rats, it was not translated to humans. These suggest that phthalate may have differential effect in disruption activities across the species. In a human study, term placentas ($n = 54$) measured altered target genes of the pathways of trophoblast differentiation and steroidogenesis (Adibi et al. 2010). Repeated measures models were used to evaluate the association of phthalate metabolites measured in third trimester urine samples with each group of target genes, accounting for correlation among the genes within a pathway. Majority of the metabolites were significantly associated with the expression of three gene targets known to be involved in trophoblast differentiation (PPARγ, AhR, HCG). Higher urinary concentrations of five phthalate metabolites were associated with lower expression of the target genes reflecting trophoblast differentiation. However, the study failed to establish associations of phthalate metabolites with the joint expression of four gene targets in the steroidgenic pathway (CYP19, 17β-HSD, P450scc, CYP1B1). Thus, phthalate and its metabolites act as endocrine-disrupting compounds to modulate placental function. The phthalate metabolites can act as agonist for PPARγ to regulate trophobalst differentiation. Further work on the phthalate exposure in the functional development of trophoblast of the placenta would be needed (Table. 9.1).

Table 9.1 Role of placenta as a biomarker for environmental exposure

Pollutants	Effects on placentation factors	References
Polychlorinated biphenyl	Alteration in the placental syncytial volume and PIGF expression	Tsuji et al. (2013)
TCDD (2,3,7,8-tetrachlorodibenzo-*p*-dioxin	Altered expression of ANG2, endoglin, and PIGF in rat placenta	Wu et al. (2014)
Persistent organic pollutants (POPs)	Increased level of PAHs and PCB in placenta correlates increased risk of neural tube defects	Ren et al. (2011)
Polychlorinated biphenyls (PCBs)	PCB exposure in utero through fish eating and environment results in LBW	Patandin et al. (1998)
Air pollution (CO, PM, NO_2)	First trimester exposure of particulate matter (2.5 μM) and carbon monoxide results in increased risk for hypertensive pregnancy	Mobasher et al. (2013)
Cadmium (Cd)	Maternal urinary Cd (0.63 mcg/L) causes LBW and altered head circumference of the girl child	Kippler et al. (2012)
Pthalate metabolism	Altered placental gene expression of trophoblast differentiation and steroidogenesis	Adibi et al. (2010)

9.6 Maternal Microbial Exposure and Its Effects on Pregnancy Outcome

Microbial colonization of the human body may begin very early through transport of maternal bacteria via the placenta (Reid et al. 2015). Microbiota differences during pregnancy are associated with serum biochemical variables of relevance to the nutritional and health status during pregnancy (cholesterol, folic acid, ferritin, and reduced transferrin) and, therefore, with possible consequences on fetal health programing (Reid et al. 2015). Studies suggested that the microbial exposure during pregnancy may influence the metabolic and immunologic profiles of the pregnant uterus and, hence, the risk of prematurity and disease developing in the offspring later in life (Vinturache et al. 2016). However, very little or no information is available whether the microbes affect the early placentation process. There was evidence for direct in utero microbial exposure and the possibility that introduction of the developing fetus to microbial products may play a role in development of postnatal immune system. Several studies suggested that maternal microbial exposure during pregnancy and perinatal and postnatal periods have a significant impact on infants' health. Further research is required on the effects of microbial exposure before conception and during gestational period that affect fetal growth. Furthermore, maternal infection with parasites and viruses during gestation has also been recently shown to modulate the immune response in the offspring. These data, along with observations from probiotic intervention studies, show that maternal microbial exposure during gestation may lead to long-term health consequences for the offspring by influencing fetal immune development.

9.7 Summary

Increasing data are now available confirming the adverse effects of environmental pollutants on human reproductive health. These chemicals can affect placentation and, thus, may have profound consequences to the developing fetus. These chemicals at lower levels can alter fundamental biological processes in placental development. Therefore, maternal exposure with these types of chemicals, at concentrations that do not cause termination of pregnancy, may cause dysfunction in pregnancy and fetal development. In conclusion, there are concerns for the environmental risk to pregnancy, indicating the need to develop more accurate tests for the protection of both the prenatal life and the development of the fetus.

References

Adibi, J. J., Whyatt, R. M., Hauser, R., Bhat, H. K., Davis, B. J., Calafat, A. M., Hoepner, L. A., Perera, F. P., Tang, D., & Williams, P. L. (2010). Transcriptional biomarkers of steroidogenesis and trophoblast differentiation in the placenta in relation to prenatal phthalate exposure. *Environmental Health Perspectives, 118*, 291–296.

Alvarez, M. M., & Chakraborty, C. (2011). Cadmium inhibits motility factor-dependent migration of human trophoblast cells. *Toxicology In Vitro, 25*, 1926–1933.

Barlow, N. J., Phillips, S. L., Wallace, D. G., Sar, M., Gaido, K. W., & Foster, P. M. (2003). Quantitative changes in gene expression in fetal rat testes following exposure to di(n-butyl) phthalate. *Toxicological Sciences, 73*, 431–441.

Jiang, Y.-Z., Wang, K., Fang, R., & Zheng, J. (2010). Expression of aryl hydrocarbon receptor in human placentas and fetal tissues. *Journal of Histochemistry and Cytochemistry, 58*, 679–685.

Kippler, M., Tofail, F., Gardner, R., Rahman, A., Hamadani, J. D., Bottai, M., & Vahter, M. (2012). Maternal cadmium exposure during pregnancy and size at birth: A prospective cohort study. *Environmental Health Perspectives, 120*, 284–289.

Mimura, J., & Fujii-Kuriyama, Y. (2003). Functional role of AhR in the expression of toxic effects by TCDD. *Biochimica et Biophysica Acta, 1619*, 263–268.

Mobasher, Z., Salam, M. T., Goodwin, T. M., Lurmann, F., Ingles, S. A., & Wilson, M. L. (2013). Associations between ambient air pollution and hypertensive disorders of pregnancy. *Environmental Research, 123*, 9–16.

Okada, Y., Meguro, M., Ohyama, H., Yoshizawa, S., Takeuchi-Hatanaka, K., Kato, N., Matsushita, S., Takashiba, S., & Nishimura, F. (2009). Human leukocyte histocompatibility antigen class II-induced cytokines from human gingival fibroblasts promote proliferation of human umbilical vein endothelial cells: potential association with enhanced angiogenesis in chronic periodontal inflammation. *Journal of Periodontal Research, 44*, 103–109.

Patandin, S., Koopman-Esseboom, C., de Ridder, M. A., Weisglas-Kuperus, N., & Sauer, P. J. (1998). Effects of environmental exposure to polychlorinated biphenyls and dioxins on birth size and growth in Dutch children. *Pediatric Research, 44*, 538–545.

Reid, G., Brigidi, P., Burton, J. P., Contractor, N., Duncan, S., Fargier, E., Hill, C., Lebeer, S., Martin, R., McBain, A. J., Mor, G., O'Neill, C., Rodriguez, J. M., Swann, J., van Hemert, S., & Ansell, J. (2015). Microbes central to human reproduction. *American Journal of Reproductive Immunology, 73*, 1–11.

Ren, A., Qiu, X., Jin, L., Ma, J., Li, Z., Zhang, L., Zhu, H., Finnell, R. H., & Zhu, T. (2011). Association of selected persistent organic pollutants in the placenta with the risk of neural tube defects. *Proceedings of the National Academy of Sciences, 108*(31), 12770–12775.

Schoeters, G. E. R., den Hond, E., Koppen, G., Smolders, R., Bloemen, K., DE Boever, P., & Govarts, E. (2011). Biomonitoring and biomarkers to unravel the risks from prenatal environmental exposures for later health outcomes. *The American Journal of Clinical Nutrition, 94*, 1964S–1969S.

Stejskalova, L., & Pavek, P. (2011). The function of cytochrome P450 1A1 enzyme (CYP1A1) and aryl hydrocarbon receptor (AhR) in the placenta. *Current Pharmaceutical Biotechnology, 12*, 715–730.

Tsuji, M., Aiko, Y., Kawamoto, T., Hachisuga, T., Kooriyama, C., Myoga, M., Tomonaga, C., Matsumura, F., Anan, A., Tanaka, M., Yu, H. S., Fujisawa, Y., Suga, R., & Shibata, E. (2013). Polychlorinated biphenyls (PCBs) decrease the placental syncytiotrophoblast volume and increase Placental Growth Factor (PlGF) in the placenta of normal pregnancy. *Placenta, 34*, 619–623.

Vinturache, A. E., Gyamfi-Bannerman, C., Hwang, J., Mysorekar, I. U., Jacobsson, B., & Preterm Birth International Collaborative. (2016). Maternal microbiome – A pathway to preterm birth. *Seminars in Fetal Neonatal Medicine, 21*(2), 94–99.

Wu, Y., Chen, X., Zhou, Q., He, Q., Kang, J., Zheng, J., Wang, K., & Duan, T. (2014). ITE and TCDD differentially regulate the vascular remodeling of rat placenta via the activation of AhR. *PLoS One, 9*, e86549.

Chapter 10
Placental Epigenetics and Its Importance in Placental Development

Contents

10.1 Introduction

Epigenetics, literally "on" genes, refers to all modifications to genes other than changes in the DNA sequence itself. Mechanisms of epigenetic inheritance include methylation of DNA, modification of histones, binding of transcription factors to chromatin, and the timing of DNA replication. Epigenetic regulation of gene expression is necessary for the correct establishment of developmental programs and for the maintenance of cell fates. Importantly, epigenetic processes are thought to be at the interface between gene regulation and the environment, such that external influences can have a major and potentially long-term impact on gene expression (Reik 2007). In fact, nutrition plays an important role in the epigenetic repertoire. Maternal diet during pregnancy is very important in fetal development but in ways that are not yet fully understood. Additionally, the maternal reproductive tract's metabolic and physiologic characteristics can also modulate the zygote's development through all embryonic stages. A strong relationship was also observed between the epigenetic alterations in the placenta and diseases of gestation and early life. These findings have critical implications for diet and other environmental factors during pregnancy (Sandovici et al. 2012).

As mentioned earlier, the major epigenetic mechanisms in mammals include DNA methylation, covalent modifications to histone proteins, RNA-mediated processes (including small RNA and long noncoding RNA), and higher-order

A.K. Duttaroy, S. Basak, *Early Nutrition and Lifestyle Factors*,
DOI 10.1007/978-3-319-38804-5_10

chromatin organization. The following sections will summarize recent studies on the role of DNA methylation and histone modifications in regulating trophoblast development and function. The roles of RNA-mediated processes in placental regulation are reviewed (Maccani and Marsit 2009).

10.2 DNA Methylation Is a Critical Regulator of Trophoblast Formation and Function

DNA methylation is a critical regulator of trophoblast formation and function. Critical period for "development window" is more susceptible where altered nutrition makes a permanent reprogramming through series of epigenetic changes at the level of epigenome alteration. Formation of trophoblast lineage and embryonic lineages happens simultaneously. Synchronization of these developments is a prerequisite for successful placentation (Fig. 10.1). The majority of DNA methylation in mammals occurs symmetrically on the cytosine base in CpG dinucleotides, termed 5-methylcytosine, and is generally associated with transcriptional repression. The mechanisms of establishment by the de novo DNA methyltransferases, Dnmt3a and Dnmt3b, and of maintenance by the DNA methyltransferase, Dnmt1 of DNA methylation, are well characterized. DNA methylation has important roles during development and regulation of genomic imprinting, X-chromosome inactivation, and preservation of chromosome stability. Genome-wide changes in DNA methylation occur during the initial stages of mammalian development, coinciding with the segregation of embryonic and trophoblast lineages (Santos et al. 2002). In mice and humans, fertilization initiates a wave of epigenetic reprogramming that is characterized by extensive DNA demethylation (Santos et al. 2002). Global DNA methylation is established at different levels in the embryonic and extraembryonic

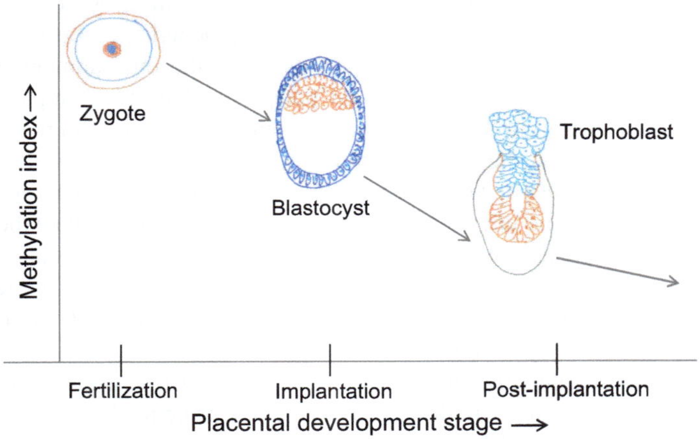

Fig. 10.1 Degree of methylation on placental development stages

lineages and creating a striking epigenetic asymmetry. These global differences in DNA methylation levels are retained in later embryo and trophoblast tissues. More recently, genome-wide analysis of mouse embryonic stem cells, and trophoblast stem cells, confirmed the asymmetry in DNA methylation levels, but suggested that the majority of the differences are likely to be within intergenic regions, repeat sequences, and centromeric heterochromatin, whereas promoter-associated regions are methylated at similar levels between embryonic and extraembryonic cells (Farthing et al. 2008). Low global levels of DNA methylation are also present in human placental tissue and purified cytotrophoblasts, thus revealing similar epigenetic asymmetries (Novakovic and Saffery 2010). Although seemingly not involved in lineage establishment, it is possible that increased levels of DNA methylation in embryonic cells are required to prevent inappropriate conversion towards a trophoblast cell fate. Consistent with this idea, there is abundant evidence to support a role for DNA methylation in lineage maintenance. Genetic inactivation of Dnmt1 results in ectopic expression of trophoblast genes in embryonic tissues at E9.5. Elf5 encodes an Ets family transcription factor that is essential for trophoblast development and appears to be epigenetically regulated (Ng et al. 2008, 2010). The Elf5 promoter is hypermethylated and transcriptionally repressed in embryonic tissues and hypomethylated and expressed in trophoblast tissues (Pearton et al. 2014). Importantly, Elf5 forms a regulatory loop with other key trophoblast factors. Therefore, when DNA methylation levels are experimentally reduced in embryonic cells, Elf5 becomes ectopically activated and cells adopt a trophoblast fate. Elf5 forms an important link between genetic and epigenetic pathways during early mouse development and suggests that epigenetically regulated factors have important roles in lineage maintenance. Importantly, DNA methylation levels are critical for regulating transcription of these genes.

Despite low levels of global DNA methylation in trophoblasts, analysis of mouse mutants has shown that DNA methylation pathways are required for placental formation and function. Genetic inactivation of Dnmt1 and Dnmt3a/Dnmt3b is lethal in mouse embryos and mutants have chorioallantoic defects (Li et al. 1992) Dnmt1-deficient placentas ectopically express the embryonic transcription factor Pou5f1, suggesting that a subset of genes need to be epigenetically repressed in trophoblast cells (Hattori et al. 2004). In addition, Dnmt3l, which is highly expressed in the chorion, is required for the formation of the labyrinth layer of the placenta (Arima et al. 2006). Mutants also reveal poor development of the spongiotrophoblast layer, suggesting a critical role for DNA methylation in regulating the specification of trophoblast lineages (Arima et al. 2006). An opportune system to study these processes further is Dnmt1/Dnmt3a/Dnmt3b mutant trophoblast stem cells, which have undetectable levels of DNA methylation at major satellite repeats. A number of imprinted genes are also affected, with elevated expression of H19 and complete repression of Igf2, which encodes a growth factor required for placental growth (Constancia et al. 2002). These changes may provide a molecular basis for certain trophoblast phenotypes associated with DNA methyltransferase mutants. Trophoblast stem cells devoid of DNA methylation are able to differentiate in vitro into a subset of trophoblast cell types including

TGC and spongiotrophoblast and could provide important molecular insight into the role of DNA methylation in specialized trophoblast cells.

10.3 Unique Features of the Placental Epigenome

Several maternal exposures have now been shown to induce epigenetic change in the offspring, often in association with altered genotype (Fig. 10.2). The most robust evidence is related to maternal smoking, linked to specific epigenetic changes in blood of offspring at birth, some of which is stable postnatally. In particular, aberrant aryl hydrocarbon receptor repressor methylation has been found in response to maternal smoking in several independent studies (please see Chap. 8). One genome-wide analysis of 790 mother–offspring dyads found a dose-dependent response of aryl hydrocarbon receptor repressor methylation in newborn cord blood. Recent findings have highlighted the role of this maternal smoking-associated epigenetic profile in mediating the effects of growth in the progeny.

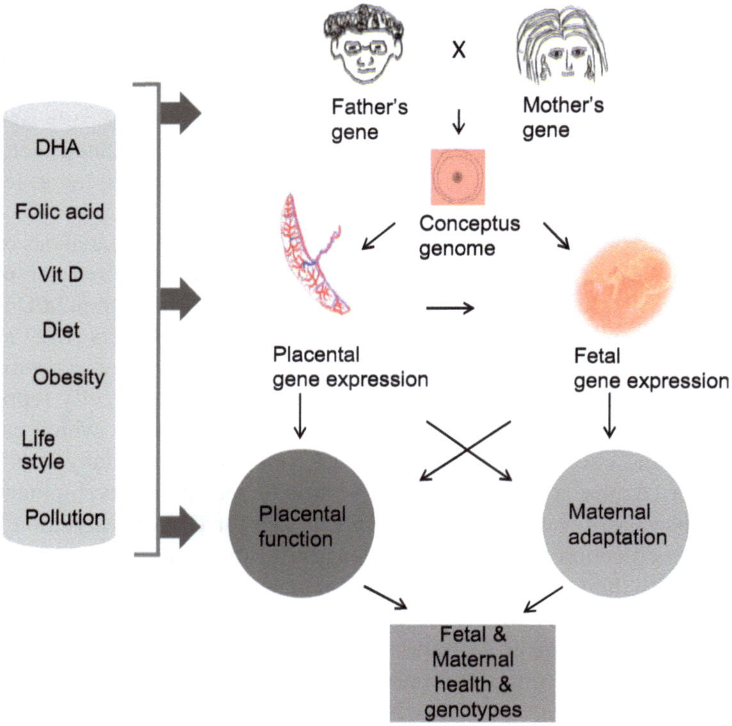

Fig. 10.2 Gene-nutrient interaction at different stages of life. Dietary, endocrine and environmental factors on human health at different levels can induce epigenetics changes that may result altered epigenomes later in life

Rather than being simply a conduit of nutrients from the maternal to fetal circulation, the placenta facilitates waste removal and plays an important role in shielding the fetus from the maternal immune system and adverse environmental exposures. The placenta is also a key endocrine organ in its own right, producing key growth factors and hormones throughout pregnancy. Being the primary interface between the mother and fetus, the placenta is susceptible to various environmental exposures that have the capacity to alter placental function and fetal development. Not surprisingly, epigenetic processes appear key to the unique functions of this important regulator of pregnancy outcome, with each placental cell lineage likely associated with specific changes in epigenetic profile.

10.4 Epigenetic Mechanisms Are Involved in Developing Placenta

Changes in DNA methylation at individual gene promoters may provide one method of transcriptional regulation during trophoblast development. As DNA methylation inhibited a reporter plasmid carrying the Ddah2 promoter sequence, these data imply that DNA methylation may regulate Ddah2 expression during trophoblast differentiation. Similar effects on individual genes may also occur during human trophoblast development. For example, reduction of DNA methylation levels by treatment of human choriocarcinoma cell lines with 5-azacytidine is associated with increased expression of E-Cadherin and loss of invasive phenotype (Rahnama et al. 2006). Trophoblast migration and invasion during development, therefore, may be mediated partly through epigenetic regulation of cell adhesion molecules. Similarly, epigenetic control of the human growth hormone locus has been well characterized in human syncytiotrophoblasts (Kimura et al. 2007). Further studies to expand understanding of these processes to a genome-wide scale will lead to a better understanding of trophoblast regulation.

Thus, overall, DNA methylation is a tightly regulated epigenetic process that is critical for trophoblast formation and function. DNA methylation can establish regulatory networks to maintain appropriate restriction of embryonic and trophoblast lineages and can modulate gene transcription and cell differentiation during trophoblast development. There is ample epidemiological and animal evidence that intrauterine and early childhood exposures affect disease and health later in life. Adverse intrauterine environments can predispose to hypertension, diabetes, and obesity in adult life. Permanent alteration in gene expression based on epigenetic modifications such as altered DNA methylation or histone modifications is a mechanism to impart a permanent memory on cells of the milieu experienced during early development. Several recent studies explored the interaction between environment and epigenetics. The DMRs differentially methylated region of certain imprinted genes are especially sensitive to environmental perturbations. The developing placenta requires the appropriate epigenetic regulation of both imprinted and non-imprinted genes. The epigenetic mechanism of DNA methylation partially regulates trophoblast differentiation and invasion into the endometrium of the uterus in order to establish and maintain a healthy placenta for the growing fetus. The placenta is an essential organ for sustaining the life of the fetus through which nutrients, oxygen exchange, immune barrier protection, and waste disposal are achieved between the fetal and maternal blood circulations (Cross et al. 1994). Defective placentation and aberrant differentiation and invasion of the trophoblasts potentially result in pathologies of pregnancy, which include PE and IUGR (Maccani and Marsit 2009; Kokkinos et al. 2010), spontaneous abortion, preterm birth (Khong and Brosens 2011), placenta accreta, and choriocarcinoma (Graham and Lala 1992; Norwitz 2006). Environmental factors, such as bad diet, smoke, oxygen stress, and pollutants, can adversely affect the epigenetic mechanisms that the embryo relies on to implant and fully grow into a healthy fetus. Epigenetic mechanisms also partially regulate EVT differentiation and invasiveness into the endometrium (Rahnama et al. 2006; Rugg-Gunn 2012; Chen and Wang 2013). Several studies demonstrated the importance of epigenetics in placental development both in human and animals. In rat placenta, drug-induced disruption of DNA methylation inhibited human invasion and proliferation of trophoblasts by disturbing the expression of epigenetically regulated genes. The deletion of placental-specific *IGF2* in mice consistently led to reduced placental growth and subsequent fetal growth restriction (Constancia et al. 2002). All these can contribute to the pathophysiology of some spontaneous miscarriages, IUGR, and preeclampsia. The association between IUGR and epigenetic variations of placental imprinted or non-imprinted genes was reported. DNA methylation pattern of 22 loci was highly predictive of IUGR. Loss of imprinting and the subsequent overexpression of the imprinted *Phlda2* gene in mice trigger placental and fetal growth retardation in the offspring, whereas its deletion caused overgrowth (Salas et al. 2004). Upregulation of *PHLDA2* in the placenta was reported in IUGR and that its expression level correlated negatively with birth weight. Other placental imprinted genes were also upregulated (*CDKN1C*) or downregulated (*MEG3*, *GATM*, *ZAC1*, *GNAS*, *MEST*, *IGF2*) in IUGR (Kobayashi 2016). Some of these differential expressions were associated with decreased placental methylation, as was the case for *H19*/*IGF2* ICR1, or loss of imprinting, as was the case for *ZAC1* and *H19* differentially methylated regions. In addition, other examples of non-imprinted genes highlight the possibility that fetal growth potential could be negatively impacted by epigenetic dysregulation in the placenta. The expression of Syncytin-1 that promotes cellular fusion in the syncytiotrophoblast was lower in human IUGR placentas. Decreased expression and aberrant DNA methylation patterns of the promoter region of the *Wnt2* gene were responsible for placental vascularization as was also observed in human pregnancy. Interestingly, epigenetic changes were also found on repeated sequences. Actually, mice with induced loss of expression of the imprinted *Cdkn1c* gene developed a preeclampsia-like syndrome, with hypertension and proteinuria. Besides, widespread DNA methylation changes were found in placentas of a cohort of patients suffering from early onset preeclampsia but not in gestational age-matched control. Some of these methylation modifications correlated negatively with expressional changes, especially for genes implicated in angiogenesis (*EPAS 1* and *FLT I*). Moreover, *BHLHE40*, a gene coding for a protein that can prevent trophoblast differentiation exhibited significantly reduced DNA methylation and increased expression in PE placentas. In addition, the expression of maspin (*SERPINB5*), a serine protease inhibitor and an inhibitor of cell migration, which may modify trophoblast cell invasion, was modified in PE (Shi et al. 2015). SERPIN A3, a critical player in implantation process, displayed increased gene expression in placentas in PE. The notion of epigenetic risk emerged in recent decades and a recent meta-analysis confirmed the increased risk of imprinting disorders such as Beckwith–Wiedemann and Silver–Russel syndromes after assisted reproductive technology (Ishida and Moore 2013). An association between increasing levels of in utero exposure to xenoestrogens, commonly found in the environment, lower placental *AluYb8* DNA methylation in boys than girls (Vilahur et al. 2014). Most studies that have examined the methylation status of imprinting genes in fetuses or placentas in animal models or in humans have associated epigenetic anomalies with adverse effects on embryonic development.

10.5 Summary

Considerable evidences suggest that maternal nutrition may have a persistent effect on the health of the offspring and may even be transmitted to the next generation. The emerging data support the hypothesis that, in addition to "thrifty genotype" inheritance, individuals with obesity, type 2 diabetes, and metabolic syndrome with an increased risk of cardiovascular diseases have suffered improper "epigenetic programming" during their fetal/postnatal development due to maternal inadequate nutrition and metabolic disturbances and that could be transmitted to the next generation(s) (Fig. 10.3). Paternal diet and behavior may also affect health of offspring via several mechanisms including epigenetics. Many environmental and lifestyle factors are linked to disrupted DNA methylation profile in the placenta affecting both global and gene-specific placental methylation, and all these may affect fetal growth and programing of health and disease in later life. The roles of epigenetic regulatory mechanisms in the placenta in fetal programming are, however, yet to be understood fully. Many environmental and lifestyle factors have now been linked to disrupted DNA methylation profiles in the placenta. Similar factors have also been linked to adverse pregnancy outcomes, leading to speculation that epigenetic variation in the placenta may be a mediator of environmental influence on pregnancy outcome. It is equally possible that epigenetic changes in the placenta observed at birth are a consequence of, and not a cause of, disrupted placental functioning and pregnancy development. It will be useful to establish variation in placental DNA methylation as the mediator of environmental determinants on placental dysfunction and pregnancy outcome. There is a wealth of data highlighting the particular role of epigenetics in placental regulation and the potential link between epigenetic dysregulation and adverse pregnancy outcomes.

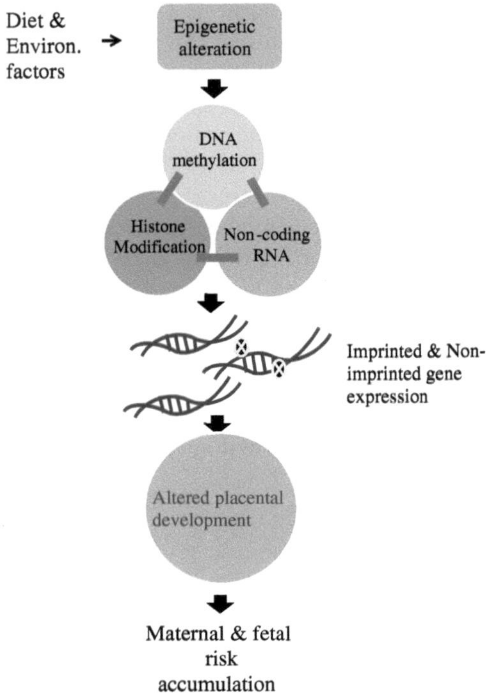

Fig. 10.3 Effects of diet and environmental factors on possible changes of the epigenome signatures of the placenta

References

Arima, T., Hata, K., Tanaka, S., Kusumi, M., Li, E., Kato, K., Shiota, K., Sasaki, H., & Wake, N. (2006). Loss of the maternal imprint in Dnmt3Lmat-/- mice leads to a differentiation defect in the extraembryonic tissue. *Developmental Biology, 297*, 361–373.

Chen, D. B., & Wang, W. (2013). Human placental microRNAs and preeclampsia. *Biology of Reproduction, 88*, 130.

Constancia, M., Hemberger, M., Hughes, J., Dean, W., Ferguson-Smith, A., Fundele, R., Stewart, F., Kelsey, G., Fowden, A., Sibley, C., & Reik, W. (2002). Placental-specific IGF-II is a major modulator of placental and fetal growth. *Nature, 417*, 945–948.

Cross, J. C., Werb, Z., & Fisher, S. J. (1994). Implantation and the placenta: Key pieces of the development puzzle. *Science, 266*, 1508–1518.

Farthing, C. R., Ficz, G., Ng, R. K., Chan, C. F., Andrews, S., Dean, W., Hemberger, M., & Reik, W. (2008). Global mapping of DNA methylation in mouse promoters reveals epigenetic reprogramming of pluripotency genes. *PLoS Genetics, 4*, e1000116.

Graham, C. H., & Lala, P. K. (1992). Mechanisms of placental invasion of the uterus and their control. *Biochemical Cell Biology, 70*, 867–874.

Hattori, N., Nishino, K., Ko, Y. G., Hattori, N., Ohgane, J., Tanaka, S., & Shiota, K. (2004). Epigenetic control of mouse Oct-4 gene expression in embryonic stem cells and trophoblast stem cells. *Journal of Biological Chemistry, 279*, 17063–17069.

Ishida, M., & Moore, G. E. (2013). The role of imprinted genes in humans. *Molecular Aspects of Medicine, 34*, 826–840.

Khong, Y., & Brosens, I. (2011). Defective deep placentation. *Best Practice and Research. Clinical Obstetrics and Gynaecology, 25*, 301–311.

Kimura, A. P., Sizova, D., Handwerger, S., Cooke, N. E., & Liebhaber, S. A. (2007). Epigenetic activation of the human growth hormone gene cluster during placental cytotrophoblast differentiation. *Molecular and Cellular Biology, 27*, 6555–6568.

Kobayashi, H. (2016). Characterization of the down-regulated genes identified in preeclampsia placenta. *Hypertens Pregnancy, 35*, 15–21.

Kokkinos, M. I., Murthi, P., Wafai, R., Thompson, E. W., & Newgreen, D. F. (2010). Cadherins in the human placenta--epithelial-mesenchymal transition (EMT) and placental development. *Placenta, 31*, 747–755.

Li, H., Subar, M., Lee, K. M., & Boussios, T. (1992). Cloning and sequence analysis of two embryonic beta-like globin cDNAs (y and z) of hamster. *Biochimica et Biophysica Acta, 1130*, 218–220.

Maccani, M. A., & Marsit, C. J. (2009). Epigenetics in the placenta. *American Journal of Reproductive Immunology, 62*, 78–89.

Ng, R. K., Dean, W., Dawson, C., Lucifero, D., Madeja, Z., Reik, W., & Hemberger, M. (2008). Epigenetic restriction of embryonic cell lineage fate by methylation of Elf5. *Nature Cell Biology, 10*, 1280–1290.

Ng, H. K., Novakovic, B., Hiendleder, S., Craig, J. M., Roberts, C. T., & Saffery, R. (2010). Distinct patterns of gene-specific methylation in mammalian placentas: Implications for placental evolution and function. *Placenta, 31*, 259–268.

Norwitz, E. R. (2006). Defective implantation and placentation: Laying the blueprint for pregnancy complications. *Reproductive Biomedicine Online, 13*, 591–599.

Novakovic, B., & Saffery, R. (2010). DNA methylation profiling highlights the unique nature of the human placental epigenome. *Epigenomics, 2*, 627–638.

Pearton, D. J., Smith, C. S., Redgate, E., van Leeuwen, J., Donnison, M., & Pfeffer, P. L. (2014). Elf5 counteracts precocious trophoblast differentiation by maintaining Sox2 and 3 and inhibiting Hand1 expression. *Developmental Biology, 392*, 344–357.

Rahnama, F., Shafiei, F., Gluckman, P. D., Mitchell, M. D., & Lobie, P. E. (2006). Epigenetic regulation of human trophoblastic cell migration and invasion. *Endocrinology, 147*, 5275–5283.

Reik, W. (2007). Stability and flexibility of epigenetic gene regulation in mammalian development. *Nature, 447*, 425–432.

Rugg-Gunn, P. J. (2012). Epigenetic features of the mouse trophoblast. *Reproductive Biomedicine Online, 25*, 21–30.

Salas, M., John, R., Saxena, A., Barton, S., Frank, D., Fitzpatrick, G., Higgins, M. J., & Tycko, B. (2004). Placental growth retardation due to loss of imprinting of Phlda2. *Mechanisms of Development, 121*, 1199–1210.

Sandovici, I., Hoelle, K., Angiolini, E., & Constancia, M. (2012). Placental adaptations to the maternal-fetal environment: Implications for fetal growth and developmental programming. *Reproductive Biomedicine Online, 25*, 68–89.

Santos, F., Hendrich, B., Reik, W., & Dean, W. (2002). Dynamic reprogramming of DNA methylation in the early mouse embryo. *Developmental Biology, 241*, 172–182.

Shi, X., Liu, H., Cao, J., Liu, Q., Tang, G., Liu, W., Liu, H., Deng, D., Qiao, F., & Wu, Y. (2015). Promoter hypomethylation of maspin inhibits migration and invasion of extravillous trophoblast cells during placentation. *PLoS One, 10*, e0135359.

Vilahur, N., Bustamante, M., Byun, H. M., Fernandez, M. F., Santa Marina, L., Basterrechea, M., Ballester, F., Murcia, M., Tardon, A., Fernandez-Somoano, A., Estivill, X., Olea, N., Sunyer, J., & Baccarelli, A. A. (2014). Prenatal exposure to mixtures of xenoestrogens and repetitive element DNA methylation changes in human placenta. *Environmental International, 71*, 81–87.

Chapter 11
Gene Regulation, microRNA, and Placentation

Contents

11.1 Introduction

Trophoblast invasion into the decidua and inner myometrium is essential for the establishment of proper implantation, maternal–fetal exchange, and immunological tolerance of the feto-placental allograft. Unlike villous trophoblasts, extravillous trophoblasts (EVT) are unique in their capacity to invade the maternal decidua and myometrium. Despite resemblance to metastasis, trophoblast invasion is tightly regulated as there are several factors involved in ensuring appropriate invasion. Like tumor cells, invasiveness is a feature of trophoblasts; however, unlike tumor invasion, trophoblast invasion is a strictly controlled physiological event. The placenta is a critical organ for the development of the fetus in utero. Trophoblast invasion into the uterine spiral arteries and the rebuilding of the arteries are crucial for normal placental development (Hunkapiller and Fisher 2008; Knofler and Pollheimer 2012). Imperfect uterine spiral artery rebuilding and inadequate trophoblast invasion have been associated with a range of pregnancy-related diseases. The placenta upregulates a variety of angiogenic factors and downregulates anti-angiogenic factors in order to stimulate angiogenesis. Spiral artery remodeling appears to be a multistage process involving spiral artery preliminary remodeling (characterized by vacuolization and apoptosis of resident endothelial cells), trophoblast migration and invasion of the decidua and myometrium, incorporation of EVTs into the spiral arteries vessel walls, and acquisition by the endovascular

A.K. Duttaroy, S. Basak, *Early Nutrition and Lifestyle Factors*,
DOI 10.1007/978-3-319-38804-5_11

EVTs of an EC-like phenotype. The trans-differentiation of EVTs into EC-like cells results in the downregulation of epithelial cell-type markers, E-cadherin and integrin α6β4, and the induction of the expression of endothelial adhesion molecules such as VE-cadherin; PECAM; NCAM; as well as integrins α5β1, α1β1, and αVβ3 (Damsky and Fisher 1998). It has been consistently reported that in early onset preeclampsia, the remodeling of the spiral arteries is impaired (Red-Horse et al. 2004). This leads to reduced perfusion of the developing placenta, subsequent intermittent hypoxia-reoxygenation episodes, reactive oxygen (ROS) production, cell injury, inflammation, and liberation of placental factors and debris into the maternal circulation.

Although a complex interplay of growth factors likely controls trophoblast cell migration and invasion, it remains unclear whether all of the currently identified effects truly play a role in vivo. There are few published reports on possible genes and microRNAs in the placentas of preeclamptic mothers. While the messenger RNAs (mRNAs) encode several target genes for these physiological events, recent research highlight the involvement of noncoding RNAs such as microRNA in feto-placental development. Among the noncoding RNAs, microRNA is the largest group that involves in gene silencing. The whole process of trophoblast invasion is similar with cancer metastasis. However, unlike cancer, the knowledge of interplay of genes and microRNAs as presented in current literature is not well established in placental biology, and some reasoning may be fully speculative.

11.2 Expression of Angiogenesis Genes in Human Placenta

Angiogenetic and anti-angiogenetic genes are important factors in placental development. Serum levels of angiogenic and anti-angiogenic factors alter with gestational age; in the same manner gene expression may change too with the progress of pregnancy. Major key players in angiogenesis are placental soluble fms-like tyrosine kinase 1 (FLT1), functioning as a VEGFR-1, collagen18 (COL18A1) functioning as an inhibitor of angiogenesis, and jagged 1 (JAG1) functioning as a signaling factor which stimulate angiogenesis in endothelial cells (Jarvenpaa et al. 2007). VEGF is a member of a family of regulatory peptides capable of controlling blood vessel formation and permeability by interacting with endothelial tyrosine kinase receptor, FLT1. FLT1 acts as a potent VEGF and placental growth factor (PGF) antagonist by binding VEGF and PGF molecules, thereby reducing free circulating concentrations of VEGF and PGF. Elevated concentrations of FLT1 are elevated in preeclamptic maternal serum as compared with control pregnancies (Maynard et al. 2003a, b, 2005, 2010). Upregulated FLT1 in placentas of pre-eclamptic mothers and excess of FLT1 concentration in early pregnancy possibly inhibit angiogenesis, contributing with other factors to placental insufficiency leading to severe pregnancy complications (Kumazaki et al. 2002). Circulating FLT1 may contribute to the pathogenesis of preeclampsia by binding circulating VEGF and PGF, thereby opposing physiological vasodilatation and contributing to the development of hypertension (Maynard et al. 2003a, b, 2005, 2010). The expression level of FLT1 was upregulated compared with normal pregnancies. FLT1 is a receptor for VEGF and PGF, which binds to its receptors with high affinity. FLT1 may function as an inhibitor of VEGF response. An increased soluble FLT1 was associated with decreased circulating levels of free VEGF and PGF, resulting in endothelial dysfunction in vitro that was rescued by endogenous VEGF and PGF (Maynard et al. 2005). Administration of FLT1 to pregnant rats induced hypertension, proteinuria, and glomerular endotheliosis, the classic lesion of preeclampsia. Excess circulating serum FLT1 concentration contributes to the pathogenesis of preeclampsia. During the last 2 months of pregnancy, the concentration of FLT1 increased and that of PGF decreased; these changes occurred earlier and were more pronounced in the women in whom preeclampsia later developed (Levine et al. 2006). The upregulation of the expression of the FLT1 gene may mean that angiogenesis is inhibited in preeclampsia. The vasculature of the feto-placental unit is highly angiogenic, and impairment in angiogenesis can lead to placental insufficiency, followed by severe pregnancy complications (Nasu et al. 2003). JAG1 and Palladin can be described as signaling genes which stimulate angiogenesis in endothelial cells. Downregulation of these genes may mean decreased angiogenesis. Downregulated ANPEP, like other aminopeptidases, participates in angiogenesis; these play a role in cell maturation, activation, or degradation, and they are involved in the local modulation of placental blood pressure.

Expression of the gene encoding COL18A1 was downregulated in preeclamptic placentas. Endostatin, the cleavage product of COL18A1, inhibits tumor angiogenesis in experimental tumor models (van Wijngaarden et al. 2004). Endostatin can also prevent endothelial cell migration (Sudhakar et al. 2003), and COL18A1 may inhibit endothelial cell proliferation. Downregulation of COL18A1 and its gene product near term may mean that the gene is active in the developmental stage, but not in late pregnancy (Jarvenpaa et al. 2007). Downregulation may possibly represent a compensatory mechanism. TNFSF12 can induce VEGF production (Hayashi et al. 2006). Downregulation of TNFSF12 expression leads to impaired VEGF function and possibly impaired vasculature. ERPIN12 works together with urokinase- or tissue-type plasminogen activators. In cooperation they degrade the extracellular matrix and help trophoblast migration. The expression profile of SERPINA3, a member of the serine proteinase inhibitor family, is also modified in preeclampsia (Chelbi et al. 2007). The expression of four genes was upregulated in the study group compared with the control group. EPAS1 is predominantly expressed in highly vascularized tissues of adult humans, in the endothelial cells of the mouse adult and embryo, and also in villus sections of the human placenta (Sood et al. 2006). Tian and associates suggested that EPAS1 may represent an important regulator of vascularization, perhaps involving the regulation of endothelial cell gene expression in response to hypoxia (Tian et al. 1997). Expression of angiogenesis-related placental genes may be altered in preeclampsia associated with IUGR. The conditions in the placenta in preeclampsia differ from those in normal placental tissue. Down- or upregulation of certain genes may only mean a normal physiological process at this stage of pregnancy.

SIGLEC10 is a member of the immunoglobulin superfamily of genes that are expressed on the cell surface. SIGLEC10 may be a possible cofactor with FLT1, but its exact function is yet to be determined. Hypoxia can lead to an increase in the expression of genes associated with by hypoxia (McCarthy et al. 2007). In addition, differences in nutritional state may impact gene expression.

Among other mechanisms of trophoblast invasion, the matrix metalloproteinase (MMP) system is of great importance. MMP2 and MMP9 play an important role in EVT invasion. Activities of MMPs are regulated by tissue inhibitors of metalloproteinases (TIMPs). For example, transforming growth factor-β1 (TGF-β1) is produced by first trimester decidual cells and limits trophoblast invasion by stimulating TIMP expression. Moreover, TNFα, produced by decidual macrophages, limits trophoblast invasion through elevation of plasminogen activator inhibitor-1 (PAI-1). Kisspeptins (KPs) have also been identified as regulators of trophoblast invasion in first trimester human trophoblast cells and in the immortalized trophoblast cell line HTR8SVneo (Hiden et al. 2007). KPs peptides are derived from the KISS-1 gene that encodes a 145 amino-acid polypeptide, that is proteolytically processed to kisspeptins of 54, 14, 13, and 10 amino acids. KP-54 (metastin) was first described as an anti-metastatic (Hiden et al. 2007). Both KP and GPR54 transcripts peak in expression in the human placenta during the first trimester of pregnancy, and thereafter their expression decreases while serum KP levels of pregnant women increase during pregnancy until term (Bilban et al. 2010). Recent studies targeted the effects of KP on the regulation of genes involved in cell invasion and angiogenesis in trophoblast primary cultures (Hiden et al. 2007). The effect of KP treatment on trophoblast migration and expression of genes involved in remodeling of the extracellular matrix and angiogenesis was studied. The different invasive trophoblast cell types produce sets of proteases, i.e., MMPs, urokinase plasminogen activator (uPA), and cathepsins, which are thought to degrade decidual extracellular matrix proteins and thereby facilitate cell invasiveness. The respective inhibitors, TIMPs and plasminogen activator inhibitors (PAIs), are produced by EVTs as well as by decidual cells to limit the extent of trophoblast invasion. Numerous soluble factors expressed at the fetal–maternal interface including chemokines, cytokines, and angiogenic factors were shown to promote trophoblast motility in an autocrine or paracrine manner (Bischof and Campana 2000; Lala and Chakraborty 2003). As a common theme, the secreted proteins were shown to stimulate MMP expression and secretion, in particular the MMP-2 and MMP-9.

11.3 Regulation of Gene Expression by Diet and Hormones

Dietary fatty acids such as EPA and DHA are considered as key components that contributes in processes of the placentation. Recent data in first trimester trophoblast demonstrated that these fatty acids directly regulate expression of angiogenic growth factors (Basak and Duttaroy 2013b) and regulate expression of lipid metabolic genes in the first trimester trophoblast cells (Johnsen et al. 2011). DHA stimulated tube formation via expression of the most potent angiogenic factor, VEGF, in these cells. DHA may also help early placentation process by increasing angiogenesis (Johnsen et al. 2011). This is in contrast with the generally observed inhibitory effects of $n-3$ LCPUFAs on angiogenesis in many cell types including tumors (Spencer et al. 2009). The mechanism responsible for the increased expression of VEGF in placental trophoblast cells by DHA is not known at present. Expression of VEGF by DHA however is very unique as its mRNA is induced by a variety of growth factors and cytokines, including PDGF (Johnsen et al. 2011), EGF, TNFα, TGF-β1, and IL-1β, but not by any fatty acids. DHA-induced VEGF expression was not accompanied with the expression of COX-2 and HIF1 genes in these cells, indicating that DHA metabolites per se may not be involved in the VEGF expression. Since the PPARγ ligand did not stimulate VEGF expression in these cells, it is unlikely that DHA stimulation of VEGF expression involves PPARγ (Gottlicher et al. 1993). DHA stimulates VEGF expression and secretion whereas fatty acids such as EPA, ARA, OA, and CLA promote ANGPTL4 secretion without affecting VEGF synthesis in the placental trophoblast cells. These data indicate the different mechanisms of action of DHA compared with other fatty acids in angiogenic process. The mechanisms that determine the angiogenic capacity of DHA compared with other fatty acids may underlie important differences in the mechanism of actions of these structurally different fatty acids. Based on available data, we postulate that these fatty acids and leptin stimulate expression of FABP4 that has been demonstrated as a target protein for VEGF. The mechanism responsible for increased secretion of ANGPTL4 in placental trophoblast cells by fatty acids other than DHA is yet to be resolved. Further work is ongoing to understand the roles of ANGPTL4, VEGF, and FABP-4 in first trimester human placental trophoblasts and their involvement in angiogenesis (Basak and Duttaroy 2013a).

11.4 MicroRNAs on Trophoblast Invasion and Angiogenesis

MicroRNA (miRNA) are endogenous small (~22 nt long) noncoding RNAs that post-transcriptionally regulate the gene expression of different proteins by binding target mRNA and inhibiting translation or inducing degradation of specific target mRNA. Although several miRNAs are ubiquitously expressed, certain miRNAs are expressed and function in a cell/tissue-specific manner, depending on the development stage or even associated with a specific physiological process. Thus, microRNA (miRNA) can regulate a variety of development and physiological processes; therefore, changes in their expression in utero due to the diets/nutrients and their effects on placentation process are not known. It could be possible that an alteration of signature molecule such as microRNA carries the altered imprint to

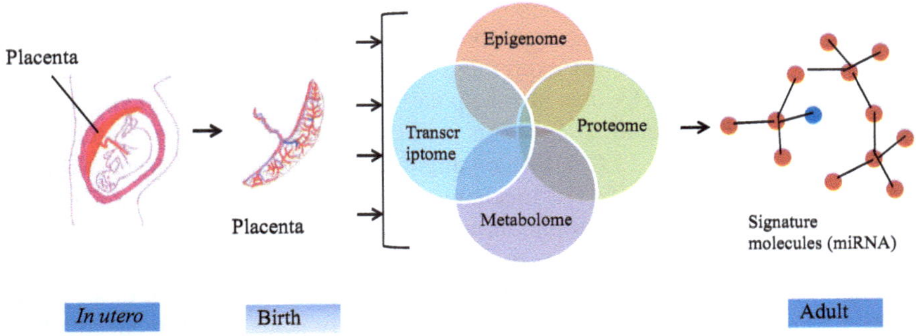

Fig. 11.1 In utero alteration of signature molecules such as microRNA carries the altered imprint to adult life and accumulates increased susceptibility of the risk factor for chronic diseases

adult life and accumulates increased susceptibility of the risk factor for chronic diseases (Fig. 11.1).

Evidences support that omega-3 fatty acids exert their physiological as well as preventive effects through the modulation of miRNA expression in cancer (Davidson et al. 2009; Vinciguerra et al. 2009), steatosis (Peyrou et al. 2010), and glioma (Farago et al. 2011). In some cases, overexpression of aopoptotic genes is reported due to reduced expression of miRNAs that target these genes. MicroRNAs (miRNAs) are the major players of trophoblast function including angiogenesis and vasculogenesis in normal and pathological conditions (Varrault et al. 2006; Renfree et al. 2013). Early in pregnancy, trophoblast invasion into the decidua and inner myometrium is essential for establishment of proper implantation, maternal–fetal exchange, and immunological tolerance of the feto-placental allograft. Unlike villous trophoblasts, extravillous trophoblasts are unique in their capacity to invade the maternal decidua and myometrium. The largest human microRNA (miRNA) gene cluster, the chromosome 19 miRNA cluster (C19MC), is expressed almost exclusively in the placenta and, rarely, in certain tumors and undifferentiated cells. The expression of C19MC miRNAs is higher in villous trophoblasts than in extravillous trophoblasts. Using a BAC-mediated overexpression of C19MC miRNAs in an extravillous trophoblast-derived cell line, C19MC miRNAs selectively attenuate cell migration without affecting cell proliferation or apoptosis. A microarray analysis revealed that C19MC miRNAs regulate target transcripts related to cellular movement. A specific C19MC member, miR-519d, directly regulates the extravillous trophoblast invasive phenotype by targeting CXCL6, NR4A2, and FOXL2 transcripts through a 3′UTR miRNA-responsive element. C19MC miRNAs modulates the migration of extravillous trophoblasts. Experimental evidence of the implication of miRNAs in the different stages of spiral arteries remodeling and their eventual dysfunction in PE is still limited. A recently published study reports an increased expression of members of the miR-17 family miRNAs (miR-17, miR-20a, and miR-20b) in PE versus normotensive human placentas (Wang et al. 2012). Among the potential targets of the miR-17 family miRNAs, the authors identified the ephrin receptor B4 (EPHB4) and ephrin B2

(EFNB2), two Eph receptor tyrosine kinases involved in angiogenesis. In the placenta, the couple EFNB2/EPHB4 directs EVTs invasion away from the placenta and towards uterine spiral arterioles. The same authors found that the forced expression of *miR-20a* in BeWo cells (a human cell line derived from a placental choriocarcinoma) inhibits syncytialization and induces the expression of ECs markers. Therefore, it has been postulated that in normal placentation, low levels of miR-17 family miRNAs allow a higher expression of EFNB2/EPHB4 in the EVTs facilitating the invasion and uterine spiral arteries remodeling. Ephrin receptors B2 and B4 are receptor tyrosine kinases, and several reports suggest their involvement in cell differentiation mechanisms. They are also involved in VEGF-mediated angiogenic functions through regulation of signaling of the VEGFR2 and VEGFR3 receptors in endothelial cells. In preeclampsia, an increased expression of miR-17 family members may suppress the expression of *EFNB2/EPHB4* resulting in inhibition of spiral arteries invasion and endovascular EVTs differentiation (Chen and Wang 2013; Doridot et al. 2013).

Recent studies have implicated the Notch pathway in the process of spiral arteries remodeling and in the pathogenesis of preeclampsia (Hunkapiller et al. 2011). The core component of this pathway consists of four transmembrane receptors (Notch1–4) and five ligands (Delta1/3/4 and Jagged 1/2) that are dynamically expressed through placental development (Hunkapiller et al. 2011). pri-miR-34a is overexpressed twice in preeclamptic placentas (where the gene promoter is demethylated), while the mature miR-34a is present at a much lower level in pathological placentas than in normal placentas.

Many other miRNAs are known to be involved in the regulation of Notch signaling. Among these, *miR-30* targets *DLL4*, and *miR-210* over-expression in ECs causes upregulation of Notch1 and induces migration and tube formation in Matrigel (Lou et al. 2012). Interestingly, miR-210 is known to be one of the most upregulated miRNAs in the preeclamptic placenta and has been involved in trophoblast migration and invasion process (Zhang et al. 2012). miR-27b regulates the expression of arterial and venous markers in human endothelium, including EFNB2, EPHB4, fms-related tyrosine kinase 1 (Flt1), and fms-related tyrosine kinase 4 (Flt) (Biyashev et al. 2012). Therefore, miRNAs regulation of Notch signaling might play an important role in placental angiogenesis. Owing to its role in endothelial progenitor cells (EPCs) physiology, a recent study suggested that miR-126 is essential for placental vasculogenesis and suggested that it could provide a therapeutic approach for PE (Yan et al. 2013). Another study has shown that *miR-29* induces apoptosis and inhibits the invasive and angiogenic capacities of trophoblast cell lines (HTR-8/SVneo, BeWo, and JAR). This study has shown that miR-29b regulates the expression of myeloid cell leukemia sequence 1 (MCL1), matrix metalloproteinase 2 (MMP2), vascular endothelial growth factor A (VEGFA), and integrin β1 (ITGB1) genes by directly binding their 3′-UTRs.

Many miRNAs have been found to play important roles in the regulation of angiogenesis and ECs homeostasis, an issue indirectly linked to trophoblast differentiation.

Despite an increasing number of studies seeking to identify miRNAs involved in the pathogenesis of PE, only a few of them have been associated to the essential process of placental angiogenesis and in particular to its earliest steps (Fu et al. 2013). In humans, the analysis of the role of miRNAs in spiral arteries remodeling is complicated due to the difficulty to access to placental tissue for experimental purposes. Therefore, it is likely that further studies will be greatly dependent on animal models (i.e., mice), but also, owing to the particularities of human placentation, on the development of in vitro models able to reproduce the different stages and cellular interactions (cell contacts, cytokines signaling, miRNA exchanges mediated by exosomes, etc.) which are likely involved in spiral artery transformation.

11.5 Summary

Placental angiogenic factors including their receptors are involved in the pathogenesis of preeclampsia. The alteration in their gene expression in early pregnancy may therefore play an important role in placenta-related disorders such as preeclampsia. Dietary lipids showed dynamic effects on placental gene expression with maternal exposure to an obesogenic diet. Nutrients such as fatty acids can modulate the expression of genes at the level of mRNA or siRNA or miRNA. It is therefore essential to understand the expression of these molecules as a part of the genome programming by means of epigenetic changes.

References

Basak, S., & Duttaroy, A. K. (2013a). cis-9, trans-11 conjugated linoleic acid stimulates expression of angiopoietin like-4 in the placental extravillous trophoblast cells. *Biochimica et Biophysica Acta, 1831*, 834–843.

Basak, S., & Duttaroy, A. K. (2013b). Effects of fatty acids on angiogenic activity in the placental extravillious trophoblast cells. *Prostaglandins, Leukotrienes, and Essential Fatty Acids, 88*, 155–162.

Bilban, M., Tauber, S., Haslinger, P., Pollheimer, J., Saleh, L., Pehamberger, H., Wagner, O., & Knöfler, M. (2010). Trophoblast invasion: Assessment of cellular models using gene expression signatures. *Placenta, 31*, 989–996.

Bischof, P., & Campana, A. (2000). A putative role for oncogenes in trophoblast invasion? *Human Reproduction, 15*(Suppl 6), 51–58.

Biyashev, D., Veliceasa, D., Topczewski, J., Topczewska, J. M., Mizgirev, I., Vinokour, E., Reddi, A. L., Licht, J. D., Revskoy, S. Y., & Volpert, O. V. (2012). miR-27b controls venous specification and tip cell fate. *Blood, 119*, 2679–2687.

Chelbi, S. T., Mondon, F., Jammes, H., Buffat, C., Mignot, T. M., Tost, J., Busato, F., Gut, I., Rebourcet, R., Laissue, P., Tsatsaris, V., Goffinet, F., Rigourd, V., Carbonne, B., Ferre, F., &

Vaiman, D. (2007). Expressional and epigenetic alterations of placental serine protease inhibitors: SERPINA3 is a potential marker of preeclampsia. *Hypertension, 49*, 76–83.

Chen, D. B., & Wang, W. (2013). Human placental microRNAs and preeclampsia. *Biology of Reproduction, 88*, 130.

Damsky, C. H., & Fisher, S. J. (1998). Trophoblast pseudo-vasculogenesis: Faking it with endothelial adhesion receptors. *Current Opinion in Cell Biology, 10*, 660–666.

Davidson, L. A., Wang, N., Shah, M. S., Lupton, J. R., Ivanov, I., & Chapkin, R. S. (2009). n-3 Polyunsaturated fatty acids modulate carcinogen-directed non-coding microRNA signatures in rat colon. *Carcinogenesis, 30*, 2077–2084.

Doridot, L., Miralles, F., Barbaux, S., & Vaiman, D. (2013). Trophoblasts, invasion, and microRNA. *Frontiers in Genetics, 4*, 248.

Farago, N., Feher, L. Z., Kitajka, K., Das, U. N., & Puskas, L. G. (2011). MicroRNA profile of polyunsaturated fatty acid treated glioma cells reveal apoptosis-specific expression changes. *Lipids in Health and Disease, 10*, 173.

Fu, S. S., Li, F. J., Wang, Y. Y., You, A. B., Qie, Y. L., Meng, X., Li, J. R., Li, B. C., Zhang, Y., & da Li, Q. (2013). Kallikrein gene-modified EPCs induce angiogenesis in rats with ischemic hindlimb and correlate with integrin alphavbeta3 expression. *PLoS One, 8*, e73035.

Gottlicher, M., Demoz, A., Svensson, D., Tollet, P., Berge, R. K., & Gustafsson, J. A. (1993). Structural and metabolic requirements for activators of the peroxisome proliferator-activated receptor. *Biochemical Pharmacology, 46*, 2177–2184.

Hayashi, T., Matsuoka, K., Saitoh, M., Takeda, S., & Kimura, M. (2006). Influence of alpha-tumor necrosis factor and beta-interleukin-1 on production of angiogenetic factors and thymidine phosphorylase activity in immortalized human decidual fibroblasts in vitro. *Journal of Obstetrics and Gynaecology Research, 32*, 15–22.

Hiden, U., Bilban, M., Knofler, M., & Desoye, G. (2007). Kisspeptins and the placenta: Regulation of trophoblast invasion. *Reviews in Endocrine & Metabolic Disorders, 8*, 31–39.

Hunkapiller, N. M., & Fisher, S. J. (2008). Chapter 12. Placental remodeling of the uterine vasculature. *Methods in Enzymology, 445*, 281–302.

Hunkapiller, N. M., Gasperowicz, M., Kapidzic, M., Plaks, V., Maltepe, E., Kitajewski, J., Cross, J. C., & Fisher, S. J. (2011). A role for Notch signaling in trophoblast endovascular invasion and in the pathogenesis of pre-eclampsia. *Development, 138*, 2987–2998.

Jarvenpaa, J., Vuoristo, J. T., Savolainen, E. R., Ukkola, O., Vaskivuo, T., & Ryynanen, M. (2007). Altered expression of angiogenesis-related placental genes in pre-eclampsia associated with intrauterine growth restriction. *Gynecological Endocrinology, 23*, 351–355.

Johnsen, G. M., Basak, S., Weedon-Fekjaer, M. S., Staff, A. C., & Duttaroy, A. K. (2011). Docosahexaenoic acid stimulates tube formation in first trimester trophoblast cells, HTR8/SVneo. *Placenta, 32*, 626–632.

Knofler, M., & Pollheimer, J. (2012). IFPA award in placentology lecture: Molecular regulation of human trophoblast invasion. *Placenta, 33*(Suppl), S55–S62.

Kumazaki, K., Nakayama, M., Suehara, N., & Wada, Y. (2002). Expression of vascular endothelial growth factor, placental growth factor, and their receptors Flt-1 and KDR in human placenta under pathologic conditions. *Human Pathology, 33*, 1069–1077.

Lala, P. K., & Chakraborty, C. (2003). Factors regulating trophoblast migration and invasiveness: Possible derangements contributing to pre-eclampsia and fetal injury. *Placenta, 24*, 575–587.

Levine, R. J., Lam, C., Qian, C., Yu, K. F., Maynard, S. E., Sachs, B. P., Sibai, B. M., Epstein, F. H., Romero, R., Thadhani, R., Karumanchi, S. A., & Group, C. S. (2006). Soluble endoglin and other circulating antiangiogenic factors in preeclampsia. *The New England Journal of Medicine, 355*, 992–1005.

Lou, Y. L., Guo, F., Liu, F., Gao, F. L., Zhang, P. Q., Niu, X., Guo, S. C., Yin, J. H., Wang, Y., & Deng, Z. F. (2012). miR-210 activates notch signaling pathway in angiogenesis induced by cerebral ischemia. *Molecular and Cellular Biochemistry, 370*, 45–51.

Maynard, S. E., Min, J. Y., Merchan, J., Lim, K. H., Li, J., Mondal, S., Libermann, T. A., Morgan, J. P., Sellke, F. W., Stillman, I. E., Epstein, F. H., Sukhatme, V. P., & Karumanchi, S. A.

(2003a). Excess placental soluble fms-like tyrosine kinase 1 (sFlt1) may contribute to endothelial dysfunction, hypertension, and proteinuria in preeclampsia. *Journal of Clinical Investigation, 111*, 649–658.

Maynard, S. E., Min, J. Y., Merchan, J., Lim, K. H., Li, J., Mondal, S., Libermann, T. A., Morgan, J. P., Sellke, F. W., Stillman, I. E., Epstein, F. H., Sukhatme, V. P., & Karumanchi, S. A. (2003b). Excess placental soluble fms-like tyrosine kinase 1 (sFlt1) may contribute to endothelial dysfunction, hypertension, and proteinuria in preeclampsia. *The Journal of Clinical Investigation, 111*, 649–658.

Maynard, S. E., Moore Simas, T. A., Bur, L., Crawford, S. L., Solitro, M. J., & Meyer, B. A. (2010). Soluble endoglin for the prediction of preeclampsia in a high risk cohort. *Hypertension in Pregnancy, 29*, 330–341.

Maynard, S. E., Venkatesha, S., Thadhani, R., & Karumanchi, S. A. (2005). Soluble Fms-like tyrosine kinase 1 and endothelial dysfunction in the pathogenesis of preeclampsia. *Pediatric Research, 57*, 1R–7R.

McCarthy, C., Cotter, F. E., Mcelwaine, S., Twomey, A., Mooney, E. E., Ryan, F., & Vaughan, J. (2007). Altered gene expression patterns in intrauterine growth restriction: Potential role of hypoxia. *American Journal of Obstetrics and Gynecology, 196*(70), e1–e6.

Nasu, K., Fujisawa, K., Nishida, Y., Kai, S., Sugano, T., Miyakawa, I., & Tateishi, Y. (2003). Expression of collagen XVIII mRNA and protein in human umbilical vein and placenta. *Reproduction, Fertility and Development, 15*, 107–114.

Peyrou, M., Bourgoin, L., & Foti, M. (2010). PTEN in non-alcoholic fatty liver disease/non-alcoholic steatohepatitis and cancer. *Digestive Diseases, 28*, 236–246.

Red-Horse, K., Zhou, Y., Genbacev, O., Prakobphol, A., Foulk, R., Mcmaster, M., & Fisher, S. J. (2004). Trophoblast differentiation during embryo implantation and formation of the maternal-fetal interface. *Journal of Clinical Investigation, 114*, 744–754.

Renfree, M. B., Suzuki, S., & Kaneko-Ishino, T. (2013). The origin and evolution of genomic imprinting and viviparity in mammals. *Philosophical Transactions of the Royal Society B-Biological Sciences, 368*, 20120151.

Sood, R., Zehnder, J. L., Druzin, M. L., & Brown, P. O. (2006). Gene expression patterns in human placenta. *Proceedings of the National Academy of Sciences of the United States of America, 103*, 5478–5483.

Spencer, L., Mann, C., Metcalfe, M., Webb, M., Pollard, C., Spencer, D., Berry, D., Steward, W., & Dennison, A. (2009). The effect of omega-3 FAs on tumour angiogenesis and their therapeutic potential. *European Journal of Cancer, 45*, 2077–2086.

Sudhakar, A., Sugimoto, H., Yang, C., Lively, J., Zeisberg, M., & Kalluri, R. (2003). Human tumstatin and human endostatin exhibit distinct antiangiogenic activities mediated by alpha v beta 3 and alpha 5 beta 1 integrins. *Proceedings of the National Academy of Sciences of the United States of America, 100*, 4766–4771.

Tian, H., McKnight, S. L., & Russell, D. W. (1997). Endothelial PAS domain protein 1 (EPAS1), a transcription factor selectively expressed in endothelial cells. *Genes and Development, 11*, 72–82.

van Wijngaarden, J., de Rooij, K., van Beek, E., Bernsen, H., Que, I., van Hinsbergh, V. W., & Lowik, C. (2004). Identification of differentially expressed genes in a renal cell carcinoma tumor model after endostatin-treatment. *Laboratory Investigation, 84*, 1472–1483.

Varrault, A., Gueydan, C., Delalbre, A., Bellmann, A., Houssami, S., Aknin, C., Severac, D., Chotard, L., Kahli, M., le Digarcher, A., Pavlidis, P., & Journot, L. (2006). Zac1 regulates an imprinted gene network critically involved in the control of embryonic growth. *Developmental Cell, 11*, 711–722.

Vinciguerra, M., Sgroi, A., Veyrat-Durebex, C., Rubbia-Brandt, L., Buhler, L. H., & Foti, M. (2009). Unsaturated fatty acids inhibit the expression of tumor suppressor phosphatase and tensin homolog (PTEN) via microRNA-21 up-regulation in hepatocytes. *Hepatology, 49*, 1176–1184.

Wang, W., Feng, L., Zhang, H., Hachy, S., Satohisa, S., Laurent, L. C., Parast, M., Zheng, J., & Chen, D. B. (2012). Preeclampsia up-regulates angiogenesis-associated microRNA (i.e., miR-17, -20a, and -20b) that target ephrin-B2 and EPHB4 in human placenta. *Journal of Clinical Endocrinology and Metabolism, 97*, E1051–E1059.

Yan, T., Liu, Y., Cui, K., Hu, B., Wang, F., & Zou, L. (2013). MicroRNA-126 regulates EPCs function: Implications for a role of miR-126 in preeclampsia. *Journal of Cellular Biochemistry, 114*, 2148–2159.

Zhang, Q., Tan, D., Luo, W., Lu, J., & Tan, Y. (2012). Expression of CD82 in human trophoblast and its role in trophoblast invasion. *PLoS One, 7*, e38487.

Chapter 12
Sources of Key Nutrients for Successful Placentation

Contents

12.1 Introduction

During the early stage of pregnancy, successful placentation depends on both fetal and maternal genes and also on the maternal nutrition status (Kind et al. 2006; Cross and Mickelson 2006; Cross et al. 1994; Brown et al. 2002). Poor maternal nutrition status in the 1st trimester of pregnancy alters the intrauterine hormonal and signaling environments and adversely affects placental growth and development. Consequently, inappropriate placental growth can affect birth weight at delivery regardless of gestational weight gain or maternal nutritional status throughout the pregnancy (Brown et al. 2002). Dutch Famine Study indicated the critical importance of maternal nutrition status in early pregnancy on overall baby's health outcome in adult life (Susser and Stein 1994). Incomplete early placentation can result in a reduced maternal supply of nutrients to the fetus that may result in poor fetal growth and leading to intrauterine growth restriction (IUGR) (Relton et al. 2005). Despite the critical importance, few data are available on the relationship between maternal nutritional status and placental growth and development (Barker 1995; Wild and Byrne 2004; Fowles et al. 2011, 2012). A prospective observational study of 3207 pregnant women in the Netherlands highlighted the

A.K. Duttaroy, S. Basak, *Early Nutrition and Lifestyle Factors*,
DOI 10.1007/978-3-319-38804-5_12

Table 12.1 Key nutrients and dietary sources

Name	Amounts per day	Source
Docosahexaenoic acid, 22:6*n* – 3	100–200 mg	Sea foods, marine fish, egg, Flaxseeds, Canola,
Folic acids	400 μg	Green leaves
Vitamin E	30 mg	Nuts, meat, eggs, cereals, tomato, sunflower seeds, nuts, spinach, safflower oil, turnip greens
Vitamin D	10 μg	Dairy products, fish and fish products
Vitamin C	500 mg	Fresh fruits
Vitamin B_{12}	2.6 μg	A daily multivitamin supplement or eat a breakfast cereal fortified with vitamin B_{12}
Magnesium	450 mg	Almonds, cucumbers cashews, mushrooms broccoli, spinach, lettuce
Zinc	11 mg	Lamb, tofu, crab meats, milk and milk products
Vitamin A	5000 IU	Liver, fish oils, milk, egg, tomato, orange
Iron	27 mg	Meat, liver, shrimp, dry fruits, grape fruits, berries, egg yolk, Sardines

association of lower feto-placental weight and maternal malnutrition (Timmermans et al. 2012). The early placentation determines the placental size and function consequently determines its nutrient transfer capacity and the growth trajectory of the fetus. Maternal undernutrition at the time of conception led to reduced birth weight and postnatal growth and the development of hypertension in experimental animals. Vitamin D and vitamin B_2, iron, folic acid, and DHA may promote, whereas *trans*-fatty acids may hamper placental angiogenesis (Fowles et al. 2011; Johnsen et al. 2011; Basak and Duttaroy 2013a, b). Maternal food consumption provides the essential nutrients (vitamin D, omega-3 and omega-6 LCPUFAs such as DHA and ARA) that may regulate physiological response to altered placental development (Fowles et al. 2011; Johnsen et al. 2011; Basak and Duttaroy 2013a, b). This chapter deals with the dietary sources of these nutrients that may favor the feto-placental development. The importance of folic acid, vitamin D, vitamin E, vitamin A, vitamin C, vitamin B_{12}, DHA, and ARA are reported in placentation processes (Wei et al. 2013; Liao et al. 2010). Key nutrients and their dietary sources are listed in Table 12.1.

12.2 Sources of Docosahexaenoic Acid, 22:6*n* – 3 (DHA) and Arachidonic Acid, 20:4*n* – 6 (ARA)

α-linolenic acid, 18:3*n* – 3 (ALA) is the precursor fatty acid of EPA and DHA. ALA and Linoleic acid, 18:2*n* – 6 (LA) are the precursors to biochemically very important LCPUFAs. The same enzymes are required for the conversion of ALA into DHA and also LA into arachidonic acid, 20:4*n* – 6 (ARA). Although it is possible for the body to convert some ALA to EPA and DHA by enlongase and desaturase enzymes, however conversion to DHA is pretty small (Neff et al. 2011). Conversion of ALA is only 0.7 % for EPA and only 0.013 % for DHA. Seafood sources such as fish and fish-oil supplements are the primary contributors of EPA and DHA in humans. However, higher intake of omega 6 fatty acids can deplete DHA from membrane phospholipids. Arachidonic acid, $20{:}4n-6$ is also very important for feto-placental growth. Moreover, both DHA and ARA levels are reduced in premature babies as they miss the placental supply of these fatty acids during the third trimester. In view of the priority given to fetal brain development, it may be that this loss is in part responsible for the major complications of prematurity in which both the brain and blood vessels are intimately involved. So a proper balance in the ratio between omega-3 and omega-6 is necessary. However, our present day food habits and preferences have given an abnormal tilt towards omega-6 with omega-3 content getting greatly reduced. This has to be corrected, and a healthy ratio of 2:1 between omega-3 and omega-6 must be aimed for our food intake. DHA is a unique nutrient that should be regularly consumed as oily sea fish or supplemented as fish oil or algal supplements. France is the only country where recommendations specifically for DHA are provided by health bodies at 120 mg for men and 100 mg for women per day. Driven by consumer demand, and despite additional production costs, modern food processors are increasingly fortifying manufactured food products, particularly milk, bread, and eggs with DHA from algae and fish oil. Meat from wild animals contains significantly higher DHA than meat from domesticated animals, because they are free-range, pasture-fed, and/or organically reared animals. Food sources of ARA are meat, seafood, and poultry. Most of the seed oils contain high concentration of LA, a precursor for synthesis of ARA in the body. ARA is very important for many biochemical processes in the feto-placental unit. However, its excess presence can precipitate many inflammatory disorders. The intake of sufficient omega-3 fats attenuates the presence of ARA. Selection of foods from various sources for balancing the presence of omega-3 and omega-6 fats will negate the harmful effects of excess of ARA.

12.3 Sources of Folic Acid

In contrast to many microorganisms and plants, humans are not able to synthesize folic acid de novo and therefore rely on its exogenous sources. Fortunately, there are many foods that are naturally rich sources of folic acid. The recommended daily intake of folic acid is 300 μg for men, 400 μg for women, and 500 μg for lactating or pregnant women. Asparagus has the highest folic acid content among the vegetables that makes it an ideal food during pregnancy. A bowl of cooked asparagus has about 79 μg of folic acid. Dietary sources of folic acid are green leafy vegetables like spinach and broccoli; beans; and fruits such as oranges, yeast, and liver (Tamura and Picciano 2006). Folic acid exists in different forms and matrixes that determine its stability and bioavailability. The folic acid shortage is very common in developing countries in poor people and women at childbearing age (Tamura and Picciano 2006). Folic acid fortification programs are in operation in many countries. Some countries have chosen, however, not to fortify with folic acid because of the potential link between high doses of synthetic folic acid and the development and progression of certain cancers avoiding neuropathy as it can mask vitamin B_{12} deficiency (Dwarkanath et al. 2013). Bio-fortification with natural folic acid produced by selected microorganisms is also used as an alternative to fortification with synthetic folic acid (Hjortmo et al. 2008).

12.4 Sources of Vitamin D, Vitamin E, and Vitamin A

Vitamin D is not truly essential in the diet as it can be synthesized in the human skin by sunlight. There are two vitamin D isoforms: cholecalciferol, vitamin D_3 (synthesized in humans) and ergocalciferol, vitamin D_2 (synthesized in some fungi and plants). Vitamin D supplements may be in the form of vitamin D_2 or D_3. The two isomers may not be equivalent in their ability to increase vitamin D status with vitamin D_3 being significantly more effective. The major circulating form of vitamin D is 25-hydroxyvitamin D3 (25(OH)D), commonly used as a marker of vitamin D status, while the biologically active form is 1,25-dihydroxyvitamin D3 (1,25(OH)2D). 25(OH)D is the precursor for synthesis of 1,25(OH)2D. The 25(OH) D is synthesised in the liver by the hydroxylation of vitamin D. A second hydroxylation reaction synthesizes 1,25(OH)2D from 25(OH)D. This takes place in the kidney but also extra-renally in many other cells and tissues including monocytes, macrophages, dendritic cells, and some epithelial cells. Inadequate intake of vitamin D has been linked with feto-placental abnormalities (Wei et al. 2013). Dietary sources of vitamin D are limited, and current concern regarding vitamin D status within the population has resulted in some countries opting to fortify foods such as milk to address this. Studies in the UK reported that around half of pregnant women are vitamin D insufficient, defined according to blood level of 25(OH)D. Currently, the UK Department of Health recommends that women consume a daily vitamin D supplement providing 10 μg (400 IU) throughout pregnancy and lactation. There is a high prevalence of vitamin D deficiency independent of whether defined as 25-hydroxyvitamin D concentrations [25(OH)D] < 50 nmol/L or <75 nmol/L. At present, dietary recommendations for vitamin D are being reevaluated in many countries. The sun exposure is the main determinant of vitamin D status at latitudes above 35°N as vitamin D synthesis is limited during the winter period. Fish and fish products are the major sources of dietary vitamin D. In addition, eggs, breads, fish oils, milk, dairy products, and mushrooms contributed to vitamin D dietary intake. Differences in assessment methods, dietary habits, and food fortification policies may account for these divergent results.

Naturally occurring vitamin E exists in eight chemical forms: (α-, β-, γ-, and δ) tocopherol and (α-, β-, γ-, and δ) tocotrienol. α-Tocopherol is the most biologically active form of vitamin E. α-Tocopherol is the only one of the eight forms of vitamin

E that contribute to meeting the vitamin E Recommended Dietary Allowance, which is 15 mg of α-tocopherol a day for adult men and women. Vegetable oils, nuts and seeds, whole grains, and green leafy vegetables are good natural sources of α-tocopherol. Some of the α-tocopherol in synthetic supplements and food additives is present as structural forms that have lower biological activity than RRR-α-tocopherol. There is little information on the effects of high doses of vitamin E consumption during human pregnancy, but no adverse effects have been noted as of yet. Vitamin E once absorbed is transported in blood by lipoproteins. Human body retains only α-tocopherol by virtue of α-Tocopherol transport protein in the liver that specifically binds α-tocopherol whereas other forms of vitamin E are lost in the bile.

Vitamin A is a fat soluble vitamin and, therefore, needs to be consumed with fat in order to have optimal absorption. The current daily value for vitamin A is 5000 international units (IU). Vitamins A and beta-carotene (a vitamin A precursor) are considered as being antioxidant nutrients and are widely found in plant or plant-derived foodstuffs. Their potential health promoting properties are well recognized, and they play a major role in affording protection in pregnancy. High vitamin A foods include sweet potatoes, carrots, dark leafy greens, winter squashes, lettuce, dried apricots, cantaloupe, bell peppers, fish, liver, and tropical fruits.

12.5 Sources of Vitamin B_{12}

Vitamin B_{12} is highly implicated in feto-placental growth and development. A slight deficiency of vitamin B_{12} can lead to anemia, fatigue, mania, and depression, while a chronic deficiency can cause a permanent damage in the brain and the nervous system. Vitamin B_{12} is only produced by certain bacteria. Human requirements for vitamin B_{12} as set by the Daily Recommended Intake (DRI) are 2–3 μg/mcg per day to upwards of 4–7 μg/mcg per day. Naturally occurring sources of vitamin B_{12} are found primarily in foods of animal origin and among fortified foods of vegetarian/vegan origin. Supplementation may be the best option for vegetarian to ensure adequate daily intake of vitamin B_{12}. The best sources of vitamin B_{12} are eggs, milk, cheese, milk products, meat, fish, shellfish, and poultry. Some soy and rice beverages as well as soy-based meat substitutes are fortified with vitamin B_{12}. The highest levels of B_{12} from vegan sources are often in the form of fortified grains, like breakfast cereals. It is best for vegans to supplement with vitamin B_{12}, as eating meat is not always a guarantee for healthy vitamin B_{12} status.

12.6 Sources of Vitamin C

Humans do not have the ability to synthesize vitamin C and therefore must depend on dietary sources to meet all requirements. The placenta can transport vitamin C by a sodium-dependent mechanism, possibly using the same transporter as for glucose. At high concentrations, the fetus increases the rate of ascorbic acid catabolism. In women taking more than 400 mg/day, this increased rate of breakdown has been linked to rare cases of neonatal scurvy. In addition to citrus fruits, many non-citrus fruits are highly rated sources, as well.

12.7 Sources of Iron, Zinc, Copper, Selenium, and Magnesium

Nutritional sources of iron are meat, poultry, fish, cereals, bread, and green vegetables. Although iron deficiency is a common cause of anemia, anemia may also result from other causes (i.e., deficiencies of folic acid, vitamin B_{12}, and vitamin B_6). Zinc is an essential trace mineral and therefore only a small amount of it (8 mg for adult women and 11 mg for adult men) is required to maintain good health. Zinc is abundantly present in meat, seafood, pulses, legumes, and whole-grain cereals. Phytates, which are commonly found in plant foods, can reduce zinc absorption, and some researchers have suggested that this increases the zinc needs of vegetarians by up to 50 %.

The richest dietary sources of copper include shellfish, nuts, seeds, legumes, grains' bran and germ, liver, and meats. Copper exhibits several biological roles being involved in connective tissues formation, iron metabolism, cardiac function, immune function, and central nervous system development. Interestingly, new insights are emerging into the role of iron and copper in neurocognitive and neurobehavioral development during the last two-thirds of gestation and in long-term consequences of their perinatal deficiency. Balances between copper and iron are recognized to assure the proper brain development, as iron deficiency results in hypo-myelination to suggest iron accumulation in rat brain during perinatal growth depends on adequate copper nutrition of dams. High magnesium containing foods include dark leafy greens, nuts, seeds, fish, beans, whole grains, avocados, yogurt, bananas, dried fruit, dark chocolate, and more.

12.8 Summary

Adequate nutrition during pregnancy is critically important both for mother and feto-placental growth and development. Nutritional needs are increased during pregnancy to support fetal growth and adaptations in maternal metabolism.

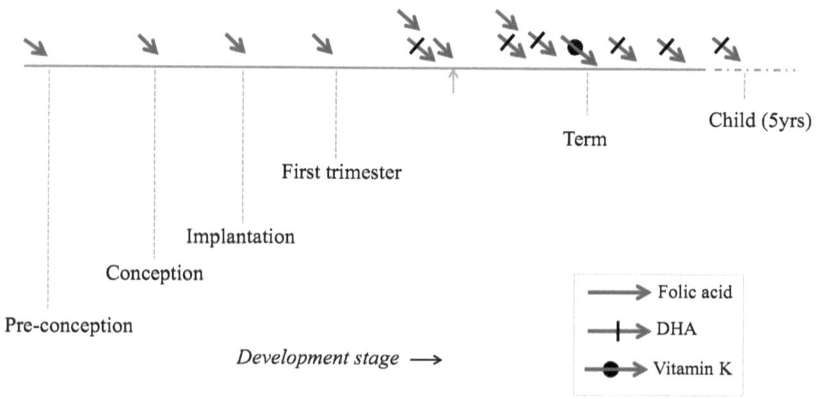

Fig. 12.1 Obligatory nutritional requirement at different stages of fetus development starts from preconception

Maternal undernutrition, caused by insufficient intakes from macronutrients and/or micronutrients, commonly exist in women from developing nations and is associated with several negative physiological outcomes in the mother and child. Although peri-conceptional use of methyl vitamins, such as folic acid, is encouraged for prevention of congenital anomalies and developmental disabilities, there are more data emerging supporting the requirements of DHA, EPA, ARA, vitamins, and minerals for feto-placental growth (Fig. 12.1). Methyl-group vitamins are intimately involved in one-carbon metabolism and regulate several metabolic processes, such as DNA methylation, gene expression, and nucleotide and monoamine synthesis. Although evidence from human trials is insufficient, results from animal models combined with the emerging recognition of the plasticity of, and the role of vitamins in, epigenetic regulation of gene expression encourage a shift of research focus toward the lasting effects of high vitamin intakes during human pregnancy to help better guide recommendations for optimal pregnancy and long-term outcomes. Maternal nutrient deficiencies increase the risk of preterm delivery, placental insufficiencies, postpartum hemorrhage and anemia, developmental birth defects, and maternal and infant death.

Modifications of the nutritional environment of the mother and fetus can also greatly impact the long-term health trajectory of the child. Vitamin consumption prior to and during pregnancy has increased as a result of proactive recommendations by health professionals, wide availability of vitamin supplements, and liberal food-fortification policies. Although few human studies have investigated the lasting effects of high vitamin intakes during pregnancy, animal models have shown that excess vitamin supplementation during gestation is associated with negative metabolic effects in both the mothers and their offspring.

References

Barker, D. J. (1995). Fetal origins of coronary heart disease. *BMJ, 311*, 171–174.

Basak, S., & Duttaroy, A. K. (2013a). cis-9, trans-11 conjugated linoleic acid stimulates expression of angiopoietin like-4 in the placental extravillous trophoblast cells. *Biochimica et Biophysica Acta, 1831*, 834–843.

Basak, S., & Duttaroy, A. K. (2013b). Effects of fatty acids on angiogenic activity in the placental extravillious trophoblast cells. *Prostaglandins, Leukotrienes, and Essential Fatty Acids, 88*, 155–162.

Brown, J. E., Murtaugh, M. A., Jacobs, D. R., Jr., & Margellos, H. C. (2002). Variation in newborn size according to pregnancy weight change by trimester. *The American Journal of Clinical Nutrition, 76*, 205–209.

Cross, J. C., & Mickelson, L. (2006). Nutritional influences on implantation and placental development. *Nutrition Reviews, 64*, S12–S18. discussion S72–S91.

Cross, J. C., Werb, Z., & Fisher, S. J. (1994). Implantation and the placenta: Key pieces of the development puzzle. *Science, 266*, 1508–1518.

Dwarkanath, P., Barzilay, J. R., Thomas, T., Thomas, A., Bhat, S., & Kurpad, A. V. (2013). High folate and low vitamin B-12 intakes during pregnancy are associated with small-for-gestational age infants in South Indian women: A prospective observational cohort study. *The American Journal of Clinical Nutrition, 98*, 1450–1458.

Fowles, E. R., Murphey, C., & Ruiz, R. J. (2011). Exploring relationships among psychosocial status, dietary quality, and measures of placental development during the first trimester in low-income women. *Biological Research for Nursing, 13*, 70–79.

Fowles, E. R., Walker, L. O., Marti, C. N., Ruiz, R. J., Wommack, J., Bryant, M., Kim, S., & Timmerman, G. M. (2012). Relationships among maternal nutrient intake and placental biomarkers during the 1st trimester in low-income women. *Archives of Gynecology and Obstetrics, 285*, 891–899.

Hjortmo, S., Patring, J., Jastrebova, J., & Andlid, T. (2008). Biofortification of folates in white wheat bread by selection of yeast strain and process. *International Journal of Food Microbiology, 127*, 32–36.

Johnsen, G. M., Basak, S., Weedon-Fekjaer, M. S., Staff, A. C., & Duttaroy, A. K. (2011). Docosahexaenoic acid stimulates tube formation in first trimester trophoblast cells, HTR8/SVneo. *Placenta, 32*, 626–632.

Kind, K. L., Moore, V. M., & Davies, M. J. (2006). Diet around conception and during pregnancy--effects on fetal and neonatal outcomes. *Reproductive Biomedicine Online, 12*, 532–541.

Liao, W. X., Wing, D. A., Geng, J. G., & Chen, D. B. (2010). Perspectives of SLIT/ROBO signaling in placental angiogenesis. *Histology and Histopathology, 25*, 1181–1190.

Neff, L. M., Culiner, J., Cunningham-Rundles, S., Seidman, C., Meehan, D., Maturi, J., Wittkowski, K. M., Levine, B., & Breslow, J. L. (2011). Algal docosahexaenoic acid affects plasma lipoprotein particle size distribution in overweight and obese adults. *Journal of Nutrition, 141*, 207–213.

Relton, C. L., Pearce, M. S., & Parker, L. (2005). The influence of erythrocyte folate and serum vitamin B12 status on birth weight. *The British Journal of Nutrition, 93*, 593–599.

Susser, M., & Stein, Z. (1994). Timing in prenatal nutrition: a reprise of the Dutch Famine Study. *Nutrition Reviews, 52*, 84–94.

Tamura, T., & Picciano, M. F. (2006). Folate and human reproduction. *The American Journal of Clinical Nutrition, 83*, 993–1016.

Timmermans, S., Steegers-Theunissen, R. P., Vujkovic, M., den Breeijen, H., Russcher, H., Lindemans, J., Mackenbach, J., Hofman, A., Lesaffre, E. E., Jaddoe, V. V., & Steegers, E. A. (2012). The Mediterranean diet and fetal size parameters: The Generation R Study. *The British Journal of Nutrition, 108*, 1399–1409.

Wei, S. Q., Audibert, F., Luo, Z. C., Nuyt, A. M., Masse, B., Julien, P., Fraser, W. D., & Group, M. S. (2013). Maternal plasma 25-hydroxyvitamin D levels, angiogenic factors, and pre-eclampsia. *American Journal of Obstetrics and Gynecology, 208*(390), e1–e6.

Wild, S. H., & Byrne, C. D. (2004). Evidence for fetal programming of obesity with a focus on putative mechanisms. *Nutrition Research Reviews, 17*, 153–162.